GALILEO'S DAUGHTER

Dava Sobel is the author of *Longitude* and co-author of *The Illustrated Longitude*. She lives in East Hampton, New York.

'*Galileo's Daughter* is like a dark old painting of those times, to which Dava Sobel holds a candle so that the crabbed figures, the over-arching church, the violent red of a cardinal's hat, are all brought to life – not to our life, not to be resurrected as if they were our contemporaries, but to their own life, bringing their own epoch with them. That is the author's skill. She writes of seventeenth-century Italy out of a full mind which has digested much reading and can light those distant days. It is an excellent book.' *Spectator*

'An entrancing historical memoir.' Lisa Jardine, *Observer*

'Once again Sobel takes her subject out of the classroom and breathes life, drama and character into these well-known historic events. An educational and entertaining read.' *Voyager*

'Pure delight.' Carol Shields, *Guardian*

'In her first book, *Longitude*, Dava Sobel demonstrated her ability to communicate complex ideas in lucid prose, and to win the reader's curiosity and sympathy by her powers as a storyteller. *Galileo's Daughter* provides the perfect vehicle for her talents. Her exemplary discussion of Galileo's research is interspersed with extracts from the letters of his daughter, which flesh out the domestic and emotional contexts to his work. Sobel's beautifully written history will ensure that Galileo's daughter is not forgotten.' *Independent on Sunday*

'A marvellous book. It brings to life this turbulent period.' *Socialist Review*

'This wonderful book blends brilliant storytelling with an elegant, modest take on the history of science. Dava Sobel's delighted respect and sympathy for the minds of the people whose lives she tells shines through all the complexities of the story and the science. And her very last, very short, sentence brings tears to the eyes.' Ruth Padel, *Independent*

'A wonderful fusion of history, science, politics and family.' *Mail on Sunday*

'Beautiful prose and haunting eloquence ... It's a wonderful book.' *Manchester Metro*

'The contrast between Galileo's life next to that of his illegitimate daughter, Maria Celeste, provides a dazzling vision of the abyss that yawned between the lives of men and women in seventeenth-century Europe. The attraction of her work lies in an ability to refract historical narrative through personal experience. Sobel does not neglect to explain, in her spare, wonderfully elegant prose, the development of Galileo's ideas and work, but she also gives us Galileo the man, Galileo the father.' Natasha Walter, *Vogue*

'From the author of *Longitude* comes a rich and unforgettable story of the relationship between the great Italian scientist and Virginia, his illegitimate daughter.' *Books Magazine*

DAVA SOBEL

GALILEO'S DAUGHTER

A Drama of Science, Faith and Love

FOURTH ESTATE • *London*

This paperback edition first published in 2000
First published in Great Britain in 1999 by
Fourth Estate Limited
6 Salem Road
London W2 4BU
www.4thestate.co.uk

7 9 10 8

Most of the quotations attributed to Galileo in these pages have been drawn
from various excellent English translations by Mary Allan-Olney, Mario
Biagioli, Richard Blackwell, Henry Crew and Alfonso de Salvio, Giorgio de
Santillana, Stillman Drake, Maurice Finocchiaro, Maria Luisa Righini
Bonelli, William Shea, Jane Sturge and Albert Van Helden. Stillman Drake
merits particular mention and gratitude for having translated every one of
Galileo's major works into English.

The Rule of Saint Clare and *The Testament of Saint Colette* (Saint Colette
solidified Saint Clare's order in France) were translated from Latin and
French texts respectively by Mother Mary Francis, Federal Abbess of Poor
Clares in America. All biblical passages are rendered from the King James
Version and from the New American Catholic Edition of the Holy Bible.
The translations of Galileo's daughter's letters from the original Italian are
the author's own.

A catalogue record for this book is available from the British Library.

ISBN 1-85702-712-4

Book design by Maura Fadden Rosenthal
Typeset by Rowland Phototypesetting Limited,
Bury St Edmunds, Suffolk
Printed and bound in Great Britain by
Clays Limited, St Ives plc

*T*o the fathers,
Galileo Galilei
&
Samuel Hillel Sobel, MD,
in loving memory

CONTENTS

PART THREE: IN ROME

PART FOUR: CARE OF THE TUSCAN EMBASSY,
VILLA MEDICI, ROME

GALILEO'S DAUGHTER

Italy in 1603

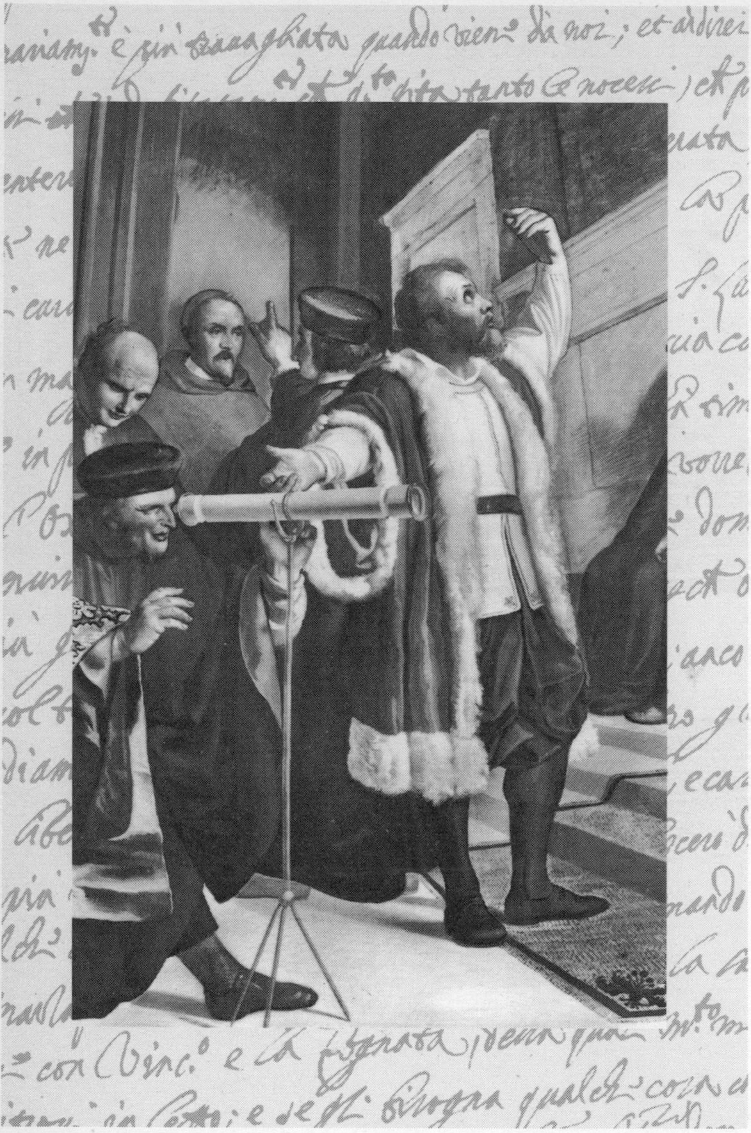

PART ONE

To Florence

[I]

She who was so precious to you

*M*OST ILLUSTRIOUS LORD FATHER

WE ARE TERRIBLY SADDENED by the death of your cherished sister, our dear aunt; but our sorrow at losing her is as nothing compared to our concern for your sake, because your suffering will be all the greater, Sire, as truly you have no one else left in your world, now that she, who could not have been more precious to you, has departed, and therefore we can only imagine how you sustain the severity of such a sudden and completely unexpected blow. And while I tell you that we share deeply in your grief, you would do well to draw even greater comfort from contemplating the general state of human misery, since we are all of us here on Earth like strangers and wayfarers, who soon will be bound for our true homeland in Heaven, where there is perfect happiness, and where we must hope that your sister's blessed soul has already gone. Thus, for the love of God, we pray you, Sire, to be consoled and to put yourself in His hands, for, as you know so well, that is what He wants of you; to do otherwise would be to injure yourself and hurt us, too, because we lament grievously when we hear that you are burdened and troubled, as we have no other source of goodness in this world but you.

I will say no more, except that with all our hearts we fervently pray the Lord to comfort you and be with you always, and we greet you dearly with our ardent love.

FROM SAN MATTEO, THE 10TH DAY OF MAY 1623.
Most affectionate daughter,

S. Maria Celeste

The day after his sister Virginia's funeral, the already world-renowned scientist Galileo Galilei received this, the first of 124 surviving letters from the once-voluminous correspondence he carried on with his elder daughter. She alone of Galileo's three children mirrored his own brilliance, industry and sensibility, and by virtue of these qualities became his confidante.

Galileo's daughter, born of his long illicit liaison with the beautiful Marina Gamba of Venice, entered the world in the summer heat of a new century, on 13 August 1600 – the same year the Dominican friar Giordano Bruno was burned at the stake in Rome for insisting, among his many heresies and blasphemies, that the Earth travelled around the Sun, instead of remaining motionless at the centre of the universe. In a world that did not yet know its place, Galileo would engage this same cosmic conflict with the Church, treading a dangerous path between the Heaven he revered as a good Catholic and the heavens he revealed through his telescope.

Galileo christened his daughter Virginia, in honour of his 'cherished sister'. But because he never married Virginia's mother, he deemed the girl herself unmarriageable. Soon after her thirteenth birthday, he placed her at the Convent of San Matteo in Arcetri, where she lived out her life in poverty and seclusion.

Virginia adopted the name Maria Celeste when she became a nun, in a gesture that acknowledged her father's fascination with the stars. Even after she professed a life of prayer and penance, she remained devoted to Galileo as though to a patron saint. The doting concern evident in her condolence letter was only to intensify over the ensuing decade as her father grew old, fell more frequently ill, pursued his singular research nevertheless, and published a book that brought him to trial by the Holy Office of the Inquisition.

The 'we' of Suor Maria Celeste's letter speaks for herself and her sister, Livia – Galileo's strange, silent second daughter, who also took the veil and vows at San Matteo to become Suor

Handmade telescope by Galileo

Arcangela. Meanwhile their brother, Vincenzio, the youngest child of Galileo and Marina's union, had been legitimised in a fiat by the grand duke of Tuscany and gone off to study law at the University of Pisa.

Thus Suor Maria Celeste consoled Galileo for being left alone in his world, with daughters cloistered in the separate world of nuns, his son not yet a man, his former mistress dead, his family of origin all deceased or dispersed.

Galileo, now fifty-nine, also stood boldly alone in his world-view, as Suor Maria Celeste knew from reading the books he wrote and the letters he shared with her from colleagues and critics all over Italy, as well as from across the continent beyond the Alps. Although her father had started his career as a professor of mathematics, teaching first at Pisa and then at Padua, every philosopher in Europe tied Galileo's name to the most startling series of astronomical discoveries ever claimed by a single individual.

In 1609, when Suor Maria Celeste was still a child in Padua, Galileo had set a telescope in the garden behind his house and turned it skywards. Never-before-seen stars leaped out of the darkness to enhance familiar constellations; the nebulous Milky Way resolved into a swath of densely packed stars; mountains and valleys pockmarked the storied perfection of the Moon; and a retinue of four attendant bodies travelled regularly around Jupiter like a planetary system in miniature.

'I render infinite thanks to God', Galileo intoned after those nights of wonder, 'for being so kind as to make me alone the first observer of marvels kept hidden in obscurity for all previous centuries.'

The new-found worlds transformed Galileo's life. He won appointment as chief mathematician and philosopher to the grand duke in 1610, and moved to Florence to assume his position at the court of Cosimo de' Medici. He took along with him his two daughters, then ten and nine years old, but he left Vincenzio, who was only four when greatness descended on the family, to live a while longer in Padua with Marina.

Galileo found himself lionised as another Columbus for his

Ptolemy's Earth-centred system of the world

conquests. Even as he attained the height of his glory, however, he attracted enmity and suspicion. For instead of opening a distant land dominated by heathens, Galileo trespassed on holy ground. Hardly had his first spate of findings stunned the populace of Europe before a new wave followed: he saw dark spots creeping continuously across the face of the Sun, and 'the mother of loves', as he called the planet Venus, cycling through phases from full to crescent, just as the Moon did.

All his observations lent credence to the unpopular Sun-centred universe of Nicolaus Copernicus, which had been introduced over half a century previously but foundered on lack of evidence. Galileo's efforts provided the beginning of a proof. And his flamboyant style of promulgating his ideas – sometimes in bawdy humorous writings, sometimes loudly at dinner parties and staged debates – transported the new astronomy from the Latin Quarters of the universities into the public arena. In 1616, a pope and a cardinal inquisitor reprimanded Galileo, warning him to curtail his forays into the supernal realms. The motions of the heavenly bodies, they said, having been touched upon in the Psalms, the

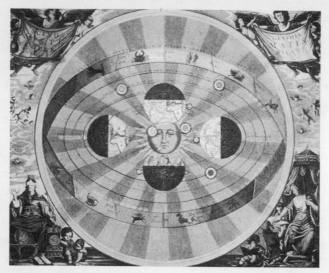

Copernicus's Sun-centred system of the world

Book of Joshua, and elsewhere in the Bible, were matters best left to the Holy Fathers of the Church.

Galileo obeyed their orders, silencing himself on the subject. For seven cautious years he turned his efforts to less perilous pursuits, such as harnessing his Jovian satellites in the service of navigation, to help sailors discover their longitude at sea. He studied poetry and wrote literary criticism. Modifying his telescope, he developed a compound microscope. 'I have observed many tiny animals with great admiration,' he reported, 'among which the flea is quite horrible, the gnat and the moth very beautiful; and with great satisfaction I have seen how flies and other little animals can walk attached to mirrors, upside down.'

Shortly after his sister's death in May of 1623, however, Galileo found reason to return to the Sun-centred universe like a moth to a flame. That summer a new pope ascended the throne of Saint Peter in Rome. The Supreme Pontiff Urban VIII brought to the Holy See an intellectualism and an interest in scientific investigation not shared by his immediate predecessors. Galileo knew the man personally – he had demonstrated his telescope to him and the two had taken the same side one night in a debate about floating bodies after a banquet at the Florentine court. Urban, for his part, had admired Galileo so long and well that he had even written a poem for him, mentioning the sights revealed by 'Galileo's glass'.

The presence of the poet pope encouraged Galileo to proceed with a long-planned popular dissertation on the two rival theories of cosmology: the Sun-centred and the Earth-centred, or, in his words, the 'two chief systems of the world'.

It might have been difficult for Suor Maria Celeste to condone this course – to reconcile her role as a bride of Christ with her father's position as potentially the greatest enemy of the Catholic Church since Martin Luther. But instead she approved of his endeavours because she knew the depth of his faith. She accepted Galileo's conviction that God had dictated the Holy Scriptures to guide men's spirits but proffered the unravelling of the universe as a challenge to their intelligence. Understanding her father's prodigious capacity in this pursuit, she prayed for his health, for his longevity, for the fulfilment of his 'every just desire'. As the convent's apothecary, she concocted elixirs and pills to strengthen him for his studies and protect him from epidemic diseases. Her letters, animated by her belief in Galileo's innocence of any heretical depravity, carried him through the ordeal of his ultimate confrontation with Urban and the Inquisition in 1633.

No detectable strife ever disturbed the affectionate relationship between Galileo and his daughter. Theirs is not a tale of abuse or rejection or intentional stifling of abilities. Rather, it is a love story, a tragedy and a mystery.

Most of Suor Maria Celeste's letters travelled in the pocket of a messenger, or in a basket laden with laundry, sweetmeats or herbal medicines, across the short distance from the Convent of San Matteo, on a hillside just south of Florence, to Galileo in the city or at his suburban home. Following the angry papal summons to Rome in 1632, however, the letters rode on horseback some two hundred miles and were frequently delayed by quarantines imposed as the Black Plague spread death and dread across Italy. Gaps of months' duration disrupt the continuity of the reportage in places, but every page is redolent of daily life, down to the pain of toothache and the smell of vinegar.

Galileo held on to his daughter's missives indiscriminately, collecting her requests for fruits or sewing supplies alongside her outbursts on ecclesiastical politics. Similarly, Suor Maria Celeste saved all of Galileo's letters, as rereading them, she often reminded him, gave her great pleasure. By the time she received the last rites, the letters she had gathered over her lifetime in the convent constituted the bulk of her earthly possessions. But then the mother abbess, who would have discovered Galileo's letters while emptying Suor Maria Celeste's cell, apparently buried or burned them out of fear. After the celebrated trial at Rome, a convent dared not harbour the writings of a 'vehemently suspected' heretic. In this fashion, the correspondence between father and daughter was long ago reduced to a monologue.

Standing in now for all the thoughts he once expressed to her are only those he chanced to offer others about her. 'A woman of exquisite mind,' Galileo described her to a colleague in another country, 'singular goodness, and most tenderly attached to me.'

On first learning of Suor Maria Celeste's letters, people generally assume that Galileo's replies must lie concealed somewhere in the recesses of the Vatican Library, and that if only an enterprising outsider could gain access, the missing half of the dialogue would be found. But, alas, the archives have been combed, several times, by religious authorities and authorised researchers all desperate to hear the paternal tone of Galileo's voice. These seekers have come to accept the account of the mother abbess's destruction of the documents as the most reasonable explanation for their disappearance. The historical importance of any paper signed by Galileo, not to mention the prices such articles have commanded for the past two centuries, leaves few conceivable places where whole packets of his letters could hide.

Although numerous commentaries, plays, poems, early lectures and manuscripts of Galileo's have also disappeared (known only by specific mentions in more than two thousand preserved letters from his contemporary correspondents), his enormous legacy includes his five most important books, two of his original hand-made telescopes, various portraits and busts he sat for during his lifetime, even parts of his body preserved after death. (The middle finger of his right hand can be seen, encased in a gilded glass egg atop an inscribed marble pedestal at the Museum of the History of Science in Florence.)

Of Suor Maria Celeste, however, only her letters remain. Bound into a single volume with cardboard and leather covers, the frayed, deckle-edged pages now reside among the rare manuscripts at Florence's National Central Library. The handwriting throughout is still legible, though the once-black ink has turned brown. Some letters bear annotations in Galileo's own hand, for he occasionally jotted notes in the margins about the things she said and at other times made seemingly unrelated calculations or geometric diagrams in the blank spaces around his address on the verso. Several of the sheets are marred by tiny holes, torn, darkened by acid or mildew, smeared with spilled oil. Of those that are water-blurred, some obviously ventured through the rain, while others look more likely tear-stained, either during the writing or the reading of them. After nearly four hundred years, the red sealing wax still sticks to the folded corners of the paper.

These letters, which have never been published in translation, recast Galileo's story. They recolour the personality and conflict of a mythic figure, whose seventeenth-century clash with Catholic doctrine continues to define the schism between science and religion. For although science has soared beyond his quaint instruments, it is still caught in his struggle, still burdened by an impression of Galileo as a renegade who scoffed at the Bible and drew fire from a Church blind to reason.

This pervasive, divisive power of the name Galileo is what Pope John Paul II tried to tame in 1992 by reinvoking his torment so long after the fact. 'A tragic mutual incomprehension', His Holiness observed of the 350-year Galileo affair, 'has been interpreted as the reflection of a fundamental opposition between science and faith.'

Yet the Galileo of Suor Maria Celeste's letters recognised no such division during his lifetime. He remained a good Catholic who believed in the power of prayer and endeavoured always to conform his duty as a scientist with the destiny of his soul. 'Whatever the course of our lives,' Galileo wrote, 'we should receive them as the highest gift from the hand of God, in which equally reposed the power to do nothing whatever for us. Indeed, we should accept misfortune not only in thanks, but in infinite gratitude to Providence, which by such means detaches us from an excessive love for Earthly things and elevates our minds to the celestial and divine.'

[II]

This grand book the universe

THE RECENTLY DECEASED RELATIVE Suor Maria Celeste mourned in her first extant letter was Virginia Galilei Landucci, the aunt she'd been named after. At the Convent of San Matteo, she shared her grief with her natural sister, Suor Arcangela (originally the namesake of Galileo's other sister, Livia), and also with her cousin Suor Chiara – the departed Virginia's own daughter Virginia.

A repetition of recollected identities echoed through the Galilei family like the sound of chanting, with its most melodic expression in the poetic rhythm of the great scientist's full name. By accepted practice among established Tuscan families in the mid-sixteenth

GALILEI GENEALOGY

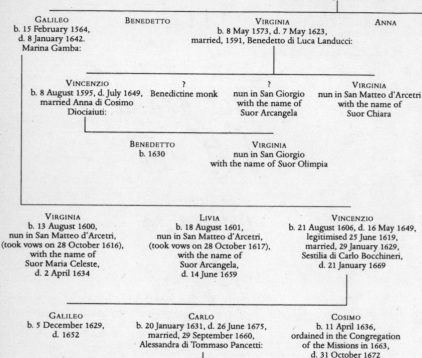

VINCENZIO
b. 1520, d. 2 July 1591,
married, 5 July 1562,
Giulia di Cosimo Ammannati
(b. 1538, d. September 1620)

GALILEO
b. 15 February 1564,
d. 8 January 1642.
Marina Gamba:

BENEDETTO

VIRGINIA
b. 8 May 1573, d. 7 May 1623,
married, 1591, Benedetto di Luca Landucci:

ANNA

VINCENZIO
b. 8 August 1595, d. July 1649,
married Anna di Cosimo
Diociaiuti:

?
Benedictine monk

?
nun in San Giorgio
with the name of
Suor Arcangela

VIRGINIA
nun in San Matteo d'Arcetri
with the name of
Suor Chiara

BENEDETTO
b. 1630

VIRGINIA
nun in San Giorgio
with the name of Suor Olimpia

VIRGINIA
b. 13 August 1600,
nun in San Matteo d'Arcetri,
(took vows on 28 October 1616),
with the name of
Suor Maria Celeste,
d. 2 April 1634

LIVIA
b. 18 August 1601,
nun in San Matteo d'Arcetri,
(took vows on 28 October 1617),
with the name of
Suor Arcangela,
d. 14 June 1659

VINCENZIO
b. 21 August 1606, d. 16 May 1649,
legitimised 25 June 1619,
married, 29 January 1629,
Sestilia di Carlo Bocchineri,
d. 21 January 1669

GALILEO
b. 5 December 1629,
d. 1652

CARLO
b. 20 January 1631, d. 26 June 1675,
married, 29 September 1660,
Alessandra di Tommaso Pancetti:

COSIMO
b. 11 April 1636,
ordained in the Congregation
of the Missions in 1663,
d. 31 October 1672

SESTILIA AND POLISSENA
b. 16 July 1662,
nuns in San Giovanni Evangelista
at San Salvi,
as of 16 January 1677,
with the names of
Suor Maria Geltrude and Suor Maria Costanza

VINCENZIO
b. 21 January 1665, d. 20 June 1709,
married, 23 December 1700,
Rosa di Niccolo Perosio (d. July 1736)

MICHELANGELO
b. 18 December 1575,
d. 3 January 1631,
married, 1608,
Anna Chiara Bandinelli (d. 1634):

LIVIA
b. 7 October 1578,
married, January 1601,
Taddeo di Cesare Galletti

LENA (?)

CESARE	?	GIROLAMO	ANTONIO
b. 15 December 1601	b. 26 September 1603, d. 27 September 1603	b. 1609	b. 1610

VINCENZIO	MECHILDE	ALBERTO CESARE	COSIMO MICHELANGELO	ELISABETTA	ANNA MARIA	MARIA FULVIA
b. 1608	d. 1634	b. November 1617, d. June 1692	d. 1634		b. 1625, d. 1634	b. 1627, d. 1634

GALILEI FAMILY COAT OF ARMS

century, when Galileo was born, the eldest son might well receive a Christian name derived from his parents' surname. Accordingly, Vincenzio Galilei and his new wife, Giulia Ammannati Galilei, attracted no special attention when they gave the name Galileo to their first child, born at Pisa on the 15th day of February in the year of Our Lord 1564. (The year was actually recorded as 1563 in the chronicles of that period, however, when New Year's Day fell on 25 March – the feast of the Annunciation.)

The family name Galilei, ironically, had itself been created from the first name of one of its foremost favourite sons. This was the renowned doctor Galileo Buonaiuti, who taught and practised medicine during the early 1400s in Florence, where he also served the government loyally. His descendants redubbed themselves the Galilei family in his honour and wrote 'Galileo Galilei' on his tombstone, but retained the coat of arms that had belonged to the ancestral Buonaiutis since the thirteenth century – a red step-ladder on a gold shield, forming a pictograph of the word *buonaiuti*, which literally means 'good help'. The meaning of the name Galileo, or Galilei, harks back to the land of Galilee, although, as Galileo explained on this score, he was not at all a Jew.

Galileo Galilei took a few tentative steps along his famous forebear's path, studying medicine for two years at the University of Pisa, before he gave himself over to the pursuit of mathematics and physics, his true passion. 'Philosophy is written in this grand book the universe, which stands continually open to our gaze,' Galileo believed. 'But the book cannot be understood unless one first learns to comprehend the language and to read the alphabet in which it is composed. It is written in the language of mathematics, and its characters are triangles, circles, and other geometric figures, without which it is humanly impossible to understand a single word of it; without these, one wanders about in a dark labyrinth.'

Galileo's father had opposed the idea of his becoming a mathematician and tried, arguing from long personal experience with mathematics and patrician poverty, to dissuade his son from choosing such a poorly paid career.

Vincenzio made a minimal living giving music lessons in the rented Pisan house where Galileo was born and partly raised. He also dabbled in the business dealings of his wife's family, the Ammannati cloth merchants, to supplement his small teaching income, but he was at heart a composer and musical theorist in the days when musical theory was considered a special branch of mathematics. Vincenzio taught Galileo to sing and to play the organ and other instruments, including the recently remodelled lute, which became their favourite. In the course of this instruction he introduced the boy to the Pythagorean rule of musical ratios, which required strict obedience in tuning and composition to numerical properties of notes in a scale. But Vincenzio subjected these prevailing rules to his own studies on the physics of sound. Music, after all, arose from vibrations in the air, not abstract concepts regarding whole numbers. Using this philosophy, Vincenzio established an ideal tuning formula for the lute by fractionally shortening the intervals between successive frets.

After Vincenzio moved to Florence with his wife in 1572, temporarily leaving Galileo behind in the care of relatives, he joined other virtuoso performers, scholars and poets bent on reviving classic Greek tragedy with music.* Vincenzio later wrote a book defending the new trend in tuning that favoured the sweetness of the instrument's sound over the ancient adherence to strict numerical relationships between notes. This book openly challenged Vincenzio's own former music teacher, who prevented its publication in Venice in 1578. Vincenzio persevered, however, until he saw the work printed in Florence three years later. None of these lessons in determination or challenge to authority was lost on the young Galileo.

'It appears to me', Vincenzio stated in his *Dialogue of Ancient and Modern Music*, 'that they who in proof of any assertion rely simply on the weight of authority, without adducing any argument in support of it, act very absurdly. I, on the contrary, wish to be

---

* Opera grew out of their efforts, officially flowering in Florence in 1600 with the first performance of *Euridice*.

to question and freely to answer you without any
...ation, as well becomes those who are in search of

...en Galileo was ten, he journeyed across Tuscany to join his
parents and his infant sister, Virginia, in Florence. He attended
grammar school near his new home until his thirteenth year, then
moved into the Benedictine monastery at Vallombrosa to take
instruction in Greek, Latin and logic. Once there, he joined the
order as a novice, hoping to become a monk himself, but his father
wouldn't let him. Vincenzio withdrew Galileo and took him
home, blaming an inflammation in the youth's eyes that required
medical attention. Money more likely decided the issue, for Vincenzio could ill afford the down payment and regular upkeep
required to support his son in a religious vocation that generated
no income. A girl was different. Vincenzio would have to pay
dowries for his daughters, either to the Church or to a husband,
and he could expect no return on either investment. Thus Vincenzio needed Galileo to grow up gainfully employed, preferably
as a doctor, so he could help support his younger sisters, now four
in number, and two brothers.

Vincenzio planned to send Galileo back to Pisa, to the College
of the Sapienza, as one of forty Tuscan boys awarded free tuition
and board, but couldn't obtain the necessary scholarship. A good
friend of Vincenzio's in Pisa offered to take Galileo into his own
home, to reduce the cost of the boy's education. Vincenzio, however, hearing that this friend was romancing one of Galileo's
Ammannati cousins, held off for three years until the love affair
ended in marriage and made the house a respectable residence for
his son.

In September of 1581, Galileo matriculated at the University of
Pisa, where medicine and mathematics both fitted into the Faculty
of Arts. Although he applied himself to the medical curriculum to
please his father, he much preferred mathematics from the moment
he encountered the geometry of Euclid in 1583. After four years of
formal study, Galileo left Pisa in 1585, at the age of twenty-one,
without completing the course requirements for a degree.

Galileo returned to his father's house in Florence. There he began behaving like a professional mathematician – writing proofs and papers in geometry, going out to give occasional public lectures, including two to the Florentine Academy on the conic configuration of Dante's Inferno, and tutoring private students. Between 1588 and 1589, when Vincenzio filled a room with weighted strings of varying lengths, diameters and tensions to test certain harmonic ideas, Galileo joined him as his assistant. It seems safe to say that Galileo, who gets credit for being the father of experimental physics, may have learned the rudiments and value of experimentation from his own father's efforts.

Having impressed several established mathematicians with his talent, Galileo procured a teaching post at the University of Pisa in 1589 and returned once more to the city of his birth at the mouth of the River Arno. The flooding of the river in fact delayed Galileo's arrival on campus, so that he missed his first six lectures and found himself fined for these absences. By the end of the year, the university authorities were docking his pay for a different sort of infraction: his refusal to wear the regulation academic regalia at all times.

Galileo deemed official doctoral dress a pretentious nuisance, and he derided the toga in a three-hundred-line verse spoof that enjoyed wide readership in that college town. Any kind of clothing got in the way of men's and women's frank appraisals of each other's attributes, he argued in ribald rhyme, while professional uniforms hid the true merits of character under a cloak of social standing. Worse, the dignity of the professor's gown barred him from the brothel, denying him the evil pleasures of whoring while resigning him to the equally sinful solace of his own hands. The gown even impeded walking, to say nothing of working.

A long black robe would surely have hindered Galileo's progress up the Leaning Tower's eight-storey spiral staircase, laden, as legend has it, with cannonballs to demonstrate a scientific principle. In that infamous episode, the weight of iron on the twenty-five-year-old professor's shoulders was as nothing compared to the burden of Aristotelian thought on his students' perceptions of reality. Not only Galileo's classes at Pisa, but university communities all over Europe, honoured the dictum of Aristotelian physics that objects of different weights fall at different speeds. A cannonball of ten pounds, for example, would be expected to fall ten times faster than a musket ball of only one pound, so that if both were released together from some summit, the cannonball would land before the musket ball had got more than one-tenth of the way to the ground. This made perfect sense to most philosophical minds, though the thought struck Galileo as preposterous. 'Try, if you can,' Galileo exhorted one of his many opponents, 'to picture in your mind the large ball striking the ground while the small one is less than a yard from the top of the tower.'

'Imagine them joining together while falling,' he appealed to another debater. 'Why should they double their speed as Aristotle claimed?' If the incongruity of these mid-air scenarios didn't deflate Aristotle's ideas, it was a simple enough matter to test his assertions with real props in a public setting.

Galileo never recorded the date or details of the actual demonstration himself but recounted the story in his old age to a young disciple, who included it in a posthumous biographical sketch. However dramatically Galileo may have executed the event, he did not succeed in swaying popular opinion down at the base of the Leaning Tower. The larger ball, being less susceptible to the effects of what Galileo recognised as air resistance, fell faster, to the great relief of the Pisan philosophy department. The fact that it fell only fractionally faster gave Galileo scant advantage.

'Aristotle says that a hundred-pound ball falling from a height of a hundred *braccia* [arm lengths] hits the ground before a one-pound ball has fallen one *braccio*. I say they arrive at the same time,' Galileo resummarised the dispute in its aftermath. 'You find, on making the test, that the larger ball beats the smaller one by two inches. Now, behind those two inches you want to hide Aristotle's ninety-nine *braccia* and, speaking only of my tiny error, remain silent about his enormous mistake.'

Indeed this was the case. Many philosophers of the sixteenth century, unaccustomed to experimental proof, much preferred the wisdom of Aristotle to the antics of Galileo, which made him an unpopular figure at Pisa.

When Vincenzio died in 1591 at the age of seventy, Galileo assumed financial responsibility for the whole family on a maths professor's meagre salary of sixty *scudi* annually. (Professors in the more venerated field of philosophy made six to eight times as much, while a father confessor could earn close to two hundred *scudi* per year, a well-trained physician about three hundred, and the commanders of the Tuscan armed forces between one thousand and two thousand five hundred.) Galileo paid out dowry instalments to his newly married sister Virginia's fractious husband, Benedetto Landucci, supported his mother and sixteen-year-old brother, Michelangelo, and maintained his sister Livia at the Convent of San Giuliano until he could arrange for her to be wed. By this time, his three other siblings had all died of childhood diseases.

Galileo lent his help ungrudgingly, even enthusiastically. 'The present I am going to make Virginia consists of a set of silken bed-hangings,' he had written home from Pisa just before her wedding. 'I bought the silk at Lucca, and had it woven, so that, though the fabric is of a wide width, it will cost me only about three *carlini* [about one-hundredth of a *scudo*] the yard. It is a striped material, and I think you will be much pleased with it. I have ordered silk fringes to match, and could very easily get the bedstead made, too. But do not say a word to anyone, that it may come to her quite unexpectedly. I will bring it when I come home for the Carnival holidays, and, as I said before, if you like I will bring her worked velvet and damask, stuff enough to make four or five handsome dresses.'

In 1592, the year after he buried his father in the Florentine church called Santa Croce, Galileo left Pisa for the chair of mathematics at the University of Padua. If he had to forsake his native Tuscany for the Serene Republic of Venice, at least he enjoyed a more distinguished position there and increased his income to 180 Venetian florins per year.

From the perspective of old age, Galileo would describe his time

The University of Padua, where Galileo taught for
eighteen years

in Padua as the happiest period of his life. He made important
friends with some of the republic's great cultural and intellectual
leaders, who invited him to their homes as well as to consult on
shipbuilding at the Venetian Arsenale. The Venetian senate
granted him a patent on an irrigation device he invented. Galileo's
influential supporters and quick-spreading reputation as an elec-
trifying lecturer earned him rises that pushed his university salary
to 300 and then to 480 florins annually. At Padua he also pursued
the seminal studies of the properties of motion that he had begun
in Pisa, for wise men regarded motion as the basis of all natural
philosophy.

Fatefully during his Paduan idyll, while visiting friends outside
the city, Galileo and two gentleman companions escaped the midday heat one afternoon by taking a siesta in an underground room.

Natural air-conditioning cooled this chamber by means of a conduit that delivered wind from a waterfall inside a nearby mountain cave. Such ingenious systems ventilated numerous sixteenth-century villas in the Italian countryside but may have admitted some noxious vapours along with the welcome zephyrs, as apparently occurred in Galileo's case. When the men awoke from their two-hour nap, they complained of various symptoms including cramps and chills, intense headache, hearing loss and muscle lethargy. Within days, the strange malaise proved fatal for one of its victims; the second man lived longer but also died of the same exposure. Galileo alone recovered. For the rest of his life, however, bouts of pain, later described by his son as arthritic or rheumatic seizures, would strike him down and confine him to his bed for weeks on end.

Under happier circumstances – although no one knows precisely when or how – Galileo in Padua met Marina Gamba, the woman who shared his private hours for twelve years and bore him three children.

Marina did not share his house, however. Galileo dwelled on Padua's Borgo dei Vignali (renamed, in recent times, Via Galileo Galilei). Like most professors, he rented out rooms to private students, many of them young noblemen from abroad, who paid to board under his roof for the duration of their private lessons with him. Marina lived in Venice, where Galileo travelled by ferry at the weekends to enjoy himself. When she became pregnant, he moved her to Padua, to a small house on the Ponte Corvo, only a five-minute walk away from his own (if one could have counted minutes in those days). Even after the ties between Marina and Galileo were strengthened by the growth of their family, their separate living arrangement remained the same.

Suor Maria Celeste Galilei, neé 'Virginia, daughter of Marina from Venice', was 'born of fornication', that is to say, out of wedlock, according to the parish registry of San Lorenzo in the city of Padua, on 13 August 1600, and baptised on the 21st. Marina was twenty-two on this occasion, and Galileo (though no mention divulges his identity), thirty-six. Such age discrepancies occurred

GALILAEUS GALILAEI PATRICIUS FLOR.
AET. SUAE
ANNUM AGENS QUADRAGESIMUM

Engraving of Galileo at age thirty-eight,
by Joseph Calendi

commonly among couples at that time. Galileo's own father had reached forty-two years before taking the twenty-four-year-old Giulia as his bride.

The following year, 1601, again in August, a registry entry on the 27th marked the baptism of 'Livia Antonia, daughter of Marina Gamba and of—' followed by a blank space.

After five more years, on 22 August 1606, a third child was baptised, 'Vincenzio Andrea, son of Madonna Marina, daughter of Andrea Gamba, and an unknown father'. Technically an 'unknown father' for not being married to the mother, Galileo nevertheless asserted his paternity by giving the baby both grandfathers' names.

Galileo recognised his illegitimate children as the heirs of his lineage, and their mother as his mate, although he ever avoided marrying Marina. Scholars by tradition tended to remain single, and the notations in the parish registry hint at circumstances that would have strengthened Galileo's resolve. After all, she was 'Marina from Venice' – not from Pisa or Florence, or Prato or Pistoia, or any other town within the bounds of Tuscany, where Galileo determined to return some day. And her heritage, 'daughter of Andrea Gamba', did not put her on a par with the poor but patrician Galilei family, whose ancestors had signed their names in the record books of a great city government.

[III]

Bright stars
speak of
your virtues

AS HIS PADUAN CAREER increased its brilliance in the early years of the seventeenth century, Galileo continued struggling to meet all his expensive family responsibilities. In 1600 his younger brother, the musical Michelangelo, was invited to play at the court of a Polish prince, and despite the maturity of his twenty-five years, he tapped Galileo for the clothing and money he needed to make the trip. Also in 1600, the same year Galileo saw the birth of his daughter Virginia, he found a husband for his sister Livia. Upon her marriage to Taddeo Galletti in 1601, Galileo negotiated the dowry, paid for the ceremony and the wedding feast, and also bought Livia's dress, which was made of black Naples velvet with light blue damask that cost more than one hundred *scudi*. Then, in 1608, Michelangelo got married, moved to Germany, and reneged on his promised share of the sisters' dowry contracts, precipitating a legal action by brother-in-law Benedetto Landucci, who complained of being cheated out of his expected sum.

Fortunately, Galileo's endeavours led him to a new source of supplemental income. In the course of teaching military architecture and fortification to private students, he had invented his first commercial scientific instrument in 1597, called the geometric and military compass. It looked like a pair of metal rulers joined by a pivot, covered all over by numbers and scales, with screws and an attachable arch to hold the compass arms open at almost any angle. By 1599, after various modifications, the device functioned as an early pocket calculator that could compute compound interest or monetary exchange rates, extract square roots for arranging armies on the battlefield, and determine the proper charge for any size of cannon. Shipwrights at the nearby Venetian Arsenale also adopted Galileo's revolutionary compass, to help them execute and test new hull designs in scale models before building them full-size.

Galileo crafted the first few compasses himself, but soon required the services of a full-time, live-in instrument maker to meet the popular demand. The hired craftsman moved into Galileo's house with wife and children in tow, to work in exchange for salary, room and board for his whole family, all production materials, and a two-thirds share of the price of the finished brass instruments, which sold for five *scudi* each. Galileo would not have made much money under these conditions, except that he charged every visiting student nearly twenty *scudi* to learn how to use the compass, and all of this was his to keep. At first he gave out a personally handwritten instruction manual as a learning aid; then in 1603 he hired an amanuensis to help generate enough copies – until three years later, when he hit on the idea of publishing the booklet for sale with the instrument.

He called his treatise *Operations of the Geometric and Military Compass of Galileo Galilei, Florentine Patrician and Teacher of*

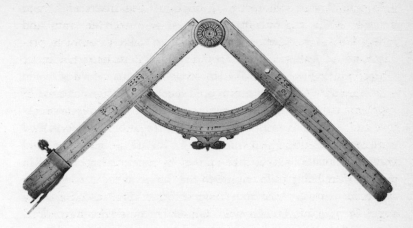

Galileo's geometric and military compass

Mathematics in the University of Padua. Its 1606 title page notes that the book was printed 'in the Author's House' and cannily dedicated to the future grand duke of Tuscany, Don Cosimo de' Medici.

'If, Most Serene Prince,' Galileo addressed his young patron in the dedication, 'I wished to set forth in this place all the praises due to your Highness's own merits and those of your distinguished family, I should be committed to such a lengthy discourse that this preface would far outrun the rest of the text, whence I shall refrain from even attempting that task, uncertain that I could finish half of it, let alone all.'

Cosimo, a lad of sixteen, had become Galileo's most elite private pupil the previous summer. The heir apparent to the House of Medici, he bore the name of his resolute grandfather, Cosimo I, who had expelled all rival and foreign influences from Florence, annexed the city of Siena to the Duchy of Tuscany, and then pressured Pope Pius V to create for him the title of grand duke in 1569. Thus the self-made Medici family, who had been successful bankers holding high government positions in the old Republic of

Medici coat of arms

Florence throughout the fourteenth and fifteenth centuries, assumed the aura and authority of royalty in Galileo's time.

Galileo, who typically returned to Florence when the University of Padua closed between terms, procured recommendations as a mathematical mentor to the royal household. As young Prince Cosimo's tutor, Galileo gained status with the boy's powerful parents: the much beloved Grand Duke Ferdinando I (who had started his career as a cardinal in Rome before being called home to the throne at the sudden death of his lecherous, murderous older brother, Francesco) and his devoutly religious French wife, Grand Duchess Cristina of Lorraine. By dedicating the tract on the geometrical compass to Cosimo, Galileo hoped to pave his way to an appointment as court mathematician – a prestigious position that would not only lighten his financial burden but also bring him home to his beloved Tuscany.

'I have waited until now to write,' Galileo said with all requisite deference in his first letter to Cosimo in 1605, 'being held back by a respectful concern of not wanting to present myself as presumptuous or arrogant. In fact, I made sure to send you the necessary signs of reverence through my closest friends and patrons, because I did not think it appropriate – leaving the darkness of the night – to appear in front of you at once and stare in the eyes of the most serene light of the rising sun without having reassured and fortified myself with their secondary and reflected rays.'

No formal contract bound the prince and the scientist at that point. If and when Galileo's tutorial services were required, he was summoned, as in the following invitation written by the chief steward of the grand duke and duchess, dated 15 August 1605, and sent from Pratolino, one of the seventeen Medici palaces, a few miles north of Florence: 'Her Most Serene Highness wishes that you should come here not only that the Prince may receive competent instruction but that your health may be restored. She hopes that the excellent air on the mountain of Pratolino will do you good. A pleasant room, good food, a comfortable bed, and a hearty welcome await you. Messer Leonido will see that you are provided with a good litter whether you wish to arrive this evening or tomorrow.'

The grand duchess again sent her horse-drawn conveyance to fetch Galileo for the wedding of Prince Cosimo, in 1608, to Maria Maddalena,

Portrait of Galileo at age forty-two, by Domenico Robusti

the archduchess of Austria and sister of Emperor Ferdinand II. The nuptials spread along both banks of the Arno, where spectators on grandstands watched a re-enactment of Jason's capture of the Golden Fleece, sumptuously staged on a specially constructed island in mid-river, with special effects including giant sea monsters that spat real fire.

In January 1609, when Grand Duke Ferdinando lay ill, Madama Cristina implored Galileo to review her husband's horoscope. Galileo's early career experience teaching astronomy to medical students had familiarised him with astrology, since doctors needed to cast horoscopes, to see what the stars foretold of patients' lives, as an aid to diagnosis and treatment, as well as to ascertain reasons for particular illnesses and determine the most propitious times for mixing medications. Galileo had prepared many horoscopes, including one for his daughter Virginia at her birth in 1600, probably for the novelty of playing with astronomical positions, as he never expressed any faith in astrological predictions. In fact he remarked how the prophecies of astrologers could most clearly be seen *after* their fulfilment.*

Nevertheless, Galileo courteously replied to the grand duchess's request by return post. Despite his forecast of many more happy years for Ferdinando, the grand duke died of his illness just three weeks later. And so it happened that Galileo's summertime student, not quite nineteen years old, was suddenly enthroned as His Serene Highness Grand Duke Cosimo II, sovereign of all Tuscany.

Cosimo's accession gave Galileo the perfect opportunity to petition for the coveted court post, as he had created it in his dreams. 'Regarding the everyday duties,' Galileo wrote in his application to Florence, 'I shun only that type of prostitution consisting of having to expose my labour to the arbitrary prices set by every customer. Instead, I will never look down on serving a prince or a great lord or those who may depend on him, but, to the contrary, I will always desire such a position.'

But he did not obtain the position just then. He continued his teaching at Padua and his research, which focused on establishing the mathematical principles of simple machines such as the lever, and determining how bodies accelerate during free fall – one of the most important unresolved questions of seventeenth-century science. 'To be ignorant of motion is to be ignorant of Nature,' Aristotle had said, and Galileo sought to end the general ignorance of Nature's laws of motion. Later that year, however, in the summer of 1609, Galileo was distracted from his motion experiments by rumours of a new Dutch curiosity called a spyglass, or eyeglass, that could make far-away objects appear closer than they were. Though few Italians had seen one first-hand, spectacle makers in Paris were already selling them in quantity.

Galileo immediately grasped the military advantage of the new spyglass, although the instrument itself, fashioned from stock spec-

* Then, as now, astrology depended on precise determinations of the positions of the wandering stars against the fixed, in order to divine the course of human events. Astronomy, which was limited in Galileo's youth to mathematical analysis of planetary motions, expanded during his lifetime to include the structure and origin of all heavenly bodies.

tacle lenses, was little more than a toy in its first incarnation. Seeking to improve the spyglass by augmenting its power, Galileo calculated the ideal shape and placement of glass, ground and polished the crucial lenses himself, and travelled to nearby Venice to show the doge, along with the entire Venetian senate, what his contrivance could do. The response, he reported, was 'the infinite amazement of all'. Even the oldest senators eagerly scaled the highest bell towers of the city, repeatedly, for the unique pleasure of discerning ships on the horizon – through the spyglass – a good two to three hours before they became visible to the keenest-sighted young lookouts.

In exchange for the gift of his telescope (as a colleague in Rome later renamed the instrument), the Venetian senate renewed Galileo's contract at the University of Padua for life, and raised his salary to one thousand florins per year – more than five times his starting pay.

Still Galileo continued to refine the optical design in subsequent attempts, and when autumn came with its early dark, he chanced to focus one of his telescopes on the face of the Moon. The jagged features that greeted him by surprise there spurred him to improve his skill at lens grinding to build even more powerful models – to revolutionise the study of astronomy by probing the actual structure of the heavens, and to disprove Aristotle's long unquestioned depiction of all celestial bodies as immutable perfect spheres.

In November 1609 Galileo fabricated lenses with double the power of the glass that had dazzled the doge. Now equipped to magnify objects by a factor of twenty, he spent half of December drafting a series of detailed drawings of the Moon in several phases. 'And it is like the face of the Earth itself,' Galileo concluded, 'which is marked here and there with chains of mountains and depths of valleys.'

From the Moon he journeyed to the stars. Two kinds of stars filled the heavens of antiquity. The 'fixed' stars outlined pictures on the night sky and wheeled around the Earth once a day. The 'wandering' stars, or planets – Mercury, Venus, Mars, Jupiter and

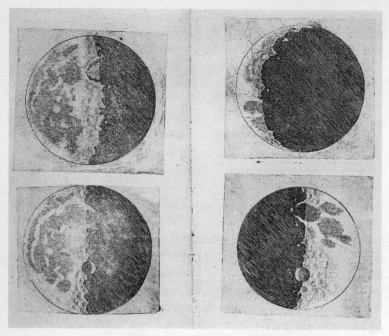

Moon drawings by Galileo in 1609

Saturn – moved against the background of the fixed stars in a complex pattern. Galileo became the first to distinguish them further: 'Planets show their globes perfectly round and definitely bounded, looking like little moons, spherical and flooded with light all over; fixed stars are never to be seen bounded by a circular periphery, but have rather the aspect of blazes whose rays vibrate about them, and they scintillate a very great deal.'*

He pursued this new nocturnal fascination through the winter, plagued by the cold and the difficulty of keeping the instrument

---

* Indeed, as any amateur astronomer today knows, stars twinkle whereas planets shine with a steady light.

steady against the trembling of his hands and the beating of his heart. He needed to wipe the lenses repeatedly with a cloth, 'or else they become fogged by the breath, humid or foggy air, or by the vapour itself which evaporates from the eye, especially when it is warm'. Early in January, he fell on the most extraordinary discovery of all: 'four planets never seen from the beginning of the world right up to our day', in orbit around the planet Jupiter.

Beyond their enormous astronomical significance, the new Jovian planets held special meaning for a friend of the Florentine court. Cosimo I of glorious memory had created a classical mythology for the Medici family when he became duke in 1537 – even before he catapulted to grand dukedom in 1569. Cosimo fashioned himself an earthly embodiment of the cosmos, as his name implied. By this coup, he convinced the Florentine citizenry that it was Medici destiny to usurp power from the other prominent families who had long governed in uneasy coalition. As the head of his *de facto* dynasty, Cosimo I identified with the planet Jupiter, named after the king of the Roman pantheon, and he filled the Palazzo della Signoria, where he lived and ruled, with frescoes stressing this Olympic theme.

Galileo had given Venice his telescope. Now he would offer Florence the moons of Jupiter.

He quickly set down his discoveries in a new book, entitled *Sidereus Nuncius*, or *The Starry Messenger*. As he had done with his earlier work on the geometric compass, he dedicated the text to young Cosimo II. On this occasion, however, Galileo took the time and gave himself enough space properly to extol his prince:

> Your Highness . . . scarcely have the immortal graces of your soul begun to shine forth on Earth than bright stars offer themselves in the heavens which, like tongues, will speak of and celebrate your most excellent virtues for all time. Behold, therefore, four stars reserved for your illustrious name, and not of the common sort and multitude of the less notable fixed stars, but of the illustrious order of wandering stars, which, indeed, make their journeys and orbits with a marvel-

lous speed around the star of Jupiter, the most noble of them all, with mutually different motions, like children of the same family, while meanwhile all together, in mutual harmony, complete their great revolutions every twelve years about the centre of the world . . .

Indeed, it appears that the Maker of the Stars himself, by clear arguments, admonished me to call these new planets by the illustrious name of Your Highness before all others. For as these stars, like the offspring worthy of Jupiter, never depart from his side except for the smallest distance, so who does not know the clemency, the gentleness of spirit, the agreeableness of manners, the splendour of the royal blood, the majesty in actions, and the breadth of authority and rule over others, all of which qualities find a domicile and exaltation for themselves in Your Highness? Who, I say, does not know that all these emanate from the most benign star of Jupiter, after God the source of all good? It was Jupiter, I say, who at Your Highness's birth, having already passed through the murky vapours of the horizon, and occupying the mid-heaven and illuminating the eastern angle from his royal house, looked down upon Your most fortunate birth from that sublime throne and poured out all his splendour and grandeur into the most pure air, so that with its first breath Your tender little body and Your soul, already decorated by God with noble ornaments, could drink in this universal power and authority.

In the continuing paean of the remaining paragraphs of this dedicatory note, Galileo took it upon himself to name the planets the Cosmian stars. But Cosimo, the eldest of eight siblings, preferred the name Medicean stars – one apiece for him and each of his three brothers. Galileo naturally bowed to this wish, though he was thus forced to paste small pieces of paper with the necessary correction over the already printed first pages in most of the 550 copies of *The Starry Messenger*.

The book created a furore. It sold out within a week of publication, so that Galileo secured only six of the thirty copies he had been promised by the printer, while news of its contents quickly spread worldwide.

Within hours after *The Starry Messenger* came off the press in Venice on 12 March 1610, the British ambassador there, Sir Henry Wotton, dispatched a copy home to King James I. 'I send herewith unto His Majesty', the ambassador wrote in his covering letter to the earl of Salisbury,

> the strangest piece of news (as I may justly call it) that he hath ever yet received from any part of the world; which is the annexed book (come abroad this very day) of the Mathematical Professor at Padua, who by the help of an optical instrument (which both enlargeth and approximateth the object) invented first in Flanders, and bettered by himself, hath discovered four new planets rolling about the sphere of Jupiter, besides many other unknown fixed stars; likewise, the true cause of the *Via Lactea* [Milky Way], so long searched; and lastly, that the moon is not spherical, but endued with many prominences, and, which is of all the strangest, illuminated with the solar light by reflection from the body of the earth, as he seemeth to say. So as upon the whole subject he hath first overthrown all former astronomy – for we must have a new sphere to save the appearances – and next all astrology. For the virtue of these new planets must needs vary the judicial part, and why may there not yet be more? These things I have been bold thus to discourse unto your Lordship, whereof here all corners are full. And the author runneth a fortune to be either exceeding famous or exceeding ridiculous. By the next ship your Lordship shall receive from me one of the above instruments, as it is bettered by this man.

In Prague, the highly respected Johannes Kepler, imperial astronomer to Rudolf II, read the emperor's copy of the book and leaped to judgment – despite the lack of a good telescope that could confirm Galileo's findings. 'I may perhaps seem rash in accepting your claims so readily with no support of my own experience,' Kepler wrote to Galileo. 'But why should I not believe a most learned mathematician, whose very style attests the soundness of his judgment?'

The copy of *The Starry Messenger* that had the greatest impact on Galileo's life, however, was the one he sent to Cosimo, along with his own superior telescope. The prince expressed his thanks late in the spring of 1610 by appointing Galileo 'Chief Mathematician of the University of Pisa and Philosopher and Mathematician to the Grand Duke'. Galileo had specifically stipulated the addition of 'Philosopher' to his title, giving himself greater prestige, but he insisted on maintaining 'Mathematician' as well, for he intended to prove the importance of mathematics in natural philosophy.

In negotiating his Tuscan future, Galileo requested the same salary he had recently been promised by the University of Padua – the figure of one thousand to be paid now in Florentine *scudi* instead of Venetian florins. Rather than plead for more money, he made the base pay stretch further by seeking official release from responsibility for his brother's share of their sisters' dowries.

Galileo also secured a bonus in personal liberty by arranging for his university appointment at Pisa to entail no noisome teaching duties. He would be free to study the world around him for the rest of his days, and to publish his discoveries for the benefit of the public under the protection of the grand duke, who promised to pay for the construction of new telescopes.

[IV]

To have the truth seen and recognised

NINE-YEAR-OLD LIVIA rode south with her father when he moved to Florence to assume his new court post in September of 1610. They left behind the serpentine canals of Venice, where the doge's palace brushed the water's edge like a fantasy spun from pink sugar and meringue. They crossed the fertile Po Valley and the Apennine spine of the Italian peninsula into the foreign country where the grand duke reigned. Italy in the seventeenth century comprised a pastiche of separate kingdoms, duchies, republics and papal states, united only by their common language, often at war with one another, and cut off from the rest of Europe by the Alps.

The landscape changed. Spires of cedar and cypress trees soared out of the rolling terrain, while ochre stucco houses sank roots into it. Here Galileo introduced Livia to the earth tones and square, sensible beauty of Tuscany. His older daughter, Virginia, already awaited them in Florence. She had gone the previous autumn at the insistence of Galileo's mother, who took Virginia home with her after an unhappy visit to Padua. Finding her son too absorbed in his new spyglass to extend the sort of hospitality she demanded, and her not-quite daughter-in-law not worthy of her attention, Madonna Giulia cut short her intended stay and returned to Tuscany.

'The little girl is so happy here', she crowed in a letter to Alessandro Piersanto, a servant in Galileo's house, 'that she will not hear that other place mentioned any more.'

Neither Virginia nor Livia had any idea when they would ever see their brother Vincenzio again. For the time being at least, Galileo deemed it best for the boy, still a toddler, to remain in Padua with Marina.

Soon after Galileo's departure, Marina married Giovanni Bartoluzzi, a respectable citizen closer to her own social station. Galileo not only approved of their union but also helped Bartoluzzi find employment with a wealthy Paduan friend of his. Still, Galileo continued sending money to Marina for Vincenzio's support, and Bartoluzzi, in turn, kept Galileo supplied with lens blanks for his telescopes, procured from the renowned glassworks on the island of Murano, within the waterways of Venice, until Florence proved a source of even better clear glass.

Galileo rented a house in Florence 'with a high terraced roof from which the whole sky is visible', where he could make his astronomical observations and install his lens-grinding lathes. While waiting for the place to become available, he stayed several months with his mother and the two little girls in rooms he let from his sister Virginia and her husband, Benedetto Landucci. Galileo's relatives provided an amicable enough atmosphere in their home, despite the recent legal fracas, but 'the malignant winter air of the city' made him miserable.

'After the absence of so many years,' Galileo lamented, 'I have experienced the very thin air of Florence as a cruel enemy of my head and the rest of my body. Colds, discharges of blood, and constipation have during the last three months reduced me to such a state of weakness, depression and despondency that I have been practically confined to the house, or rather to my bed, but without the blessing of sleep or rest.'

He devoted what time his health allowed to the problem of Saturn, much further away than Jupiter – at the apparent limit of his best telescope's resolution – where he thought he could just discern two large, immobile moons. He described what he had seen in a Latin anagram, which, when correctly unscrambled, said, 'I observed the highest planet to be triple-bodied.' Thus staking his claim to the new discovery without making a fool of himself before establishing proper confirmation, he dispatched the anagram to several well-known astronomers. None of them correctly decoded it, however. The great Kepler in Prague, who by this point had held the telescope and deemed it 'more precious than any sceptre', misinterpreted the message to mean Galileo had discovered two moons at Mars.*

All through that same autumn of 1610, with Venus visible in the evening sky, Galileo studied the planet's changing size and shape. He kept a telescope trained on Jupiter, too, in a protracted struggle to ascertain the precise orbital periods of the four new satellites further to validate their reality. Meanwhile, other astronomers complained of struggling just to catch sight of the Jovian satellites through inferior instruments, and therefore they questioned the bodies' very existence. Despite Kepler's endorsement, some sniped that the moons must be optical illusions, suspiciously introduced into the sky by Galileo's lenses.

Now that the moons had become matters of the Florentine

---

* Although Kepler erred here, two moons did turn up in telescopic views of Mars more than two centuries later, when Asaph Hall at the US Naval Observatory detected the Martian satellites he named Phobos and Deimos.

state, this situation required immediate remedy to protect the honour of the grand duke. Galileo scrambled to build as many telescopes as he could for export to France, Spain, England, Poland, Austria, as well as for princes all around Italy. 'In order to maintain and increase the renown of these discoveries,' he reasoned, 'it appears to me necessary . . . to have the truth seen and recognised, by means of the effect itself, by as many people as possible.'

Famous philosophers, including some of Galileo's former colleagues at Pisa, refused to look through any telescope at the purported new contents of Aristotle's immutable cosmos. Galileo deflected their slurs with humour: learning of the death of one such opponent in December 1610, he wished aloud that the professor, having ignored the Medicean stars during his time on Earth, might now encounter them *en route* to Heaven.

To cement the primacy of his claims, Galileo thought it politic to visit Rome and publicise his discoveries around the Eternal City. He had travelled there once before, in 1587, to discuss geometry with the pre-eminent Jesuit mathematician Christoph Clavius, who had written influential commentaries on astronomy, and who now would surely welcome news of Galileo's recent work. Grand Duke Cosimo condoned the trip. He thought it might heighten his own stature in Rome, where his brother Carlo currently filled the traditional position of resident Medici cardinal.

Unfortunately, Galileo's sickly reaction to the air of Florence prevented him from setting out until 23 March 1611. He spent six days on the road in the grand duke's litter, and at night he set up his telescope in every stop along the way – San Casciano, Siena, San Quirico, Acquapendente, Viterbo, Alonterosi – to continue tracking the revolutions of Jupiter's moons.

Upon Galileo's arrival at the week's end, the warmth of his Roman welcome surprised him. 'I have been received and fêted by many illustrious cardinals, prelates and princes of this city,' he reported, 'who wanted to see the things I have observed and were much pleased, as I was too on my part in viewing the marvels of their statuary, paintings, frescoed rooms, palaces, gardens, etc.'

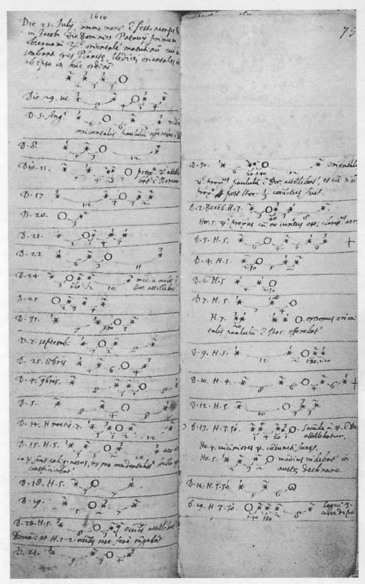

Page from Galileo's notebook tracking the orbits of the
satellites of Jupiter

The Collegio Romano

Galileo garnered the powerful endorsement of the Collegio Romano, the central institution of the Jesuit educational network, where Father Clavius, now well into his seventies, was chief mathematician. He and his revered colleagues, regarded by the Church as the top astronomical authorities, had obtained telescopes of their own, and now as a group corroborated all of Galileo's observations. Bound as these Jesuits were to Aristotelian belief in an unchanging cosmos, they did not deny the evidence of their senses. They even honoured Galileo with a rare invitation to visit.

'On Friday evening of the past week in the Collegio Romano,' a social bulletin reported in early April, 'in the presence of cardinals and of the Marquis of Monticelli, its promoter, a Latin oration was recited, with other compositions in praise of Signor Galileo Galilei, mathematician to the grand duke, magnifying and exalting to the heavens his new observation of new planets that were unknown to the ancient philosophers.'

This marquis of Monticelli who attended Galileo's fête was an affable, idealistic young Roman named Federico Cesi. His handful of noble titles also pronounced him duke of Acquasparta and prince of San Polo and Sant'Angelo. In addition to these honours he bore by birth, he had distinguished himself in 1603, at the age of eighteen, by founding the world's first scientific society, the Lyncean Academy. Cesi pooled his wealth, foresight and curiosity to establish a forum free from university control or prejudice. He made the academy international from the outset – one of its four charter members being Dutch – and multidisciplinary by design: 'The Lyncean Academy desires as its members philosophers who are eager for real knowledge and will give themselves to the study of nature, especially mathematics; at the same time it will not neglect the ornaments of elegant literature and philology, which, like graceful garments, adorn the whole body of science.' The choice of the sharp-eyed lynx as totem emphasised the importance Cesi placed on faithful observation of Nature. At official ceremonies, Cesi sometimes wore a lynx pendant on a gold neck chain.

Cesi entreated Galileo, who embodied the Lynceans' organising principles, to join the academy during his stay in Rome. He held a banquet in Galileo's honour on 14 April on the city's highest hill, where one of the other dinner guests, Greek mathematician Giovanni Demisiani, proposed the name 'telescope' for the spyglass Galileo had brought along to show the party the moons of Jupiter. The men lingered long into the night enjoying the novel views. To dispel any possible doubt about his instrument's veracity, Galileo also aimed the telescope point-blank at the exterior wall of the Lateran Church, where a chiselled inscription attributed to Pope Sixtus V could be easily read by all, though it stood over a mile away.

Galileo's formal election to the Lyncean Academy the next week privileged him to add the title 'Lyncean' after his signature on any literary work or private correspondence, which practice he took up immediately. Furthermore the academy, Cesi promised, would become Galileo's publisher.

Before leaving Rome triumphant at the end of May, Galileo gained a favourable audience with the reigning pope, Paul V, who ordinarily took no great interest in science or scientists. Galileo also made the acquaintance of Maffeo Cardinal Barberini, the man destined to become the future Pope Urban VIII. Cardinal Barberini, a fellow Tuscan roughly the same age as Galileo and, like him, an alumnus of the University of Pisa, admired the court philosopher's scientific work and shared his interest in poetry.

Chance threw Galileo and Barberini together again the following autumn, in Florence, when the visiting cardinal was the grand duke's dinner guest, and Galileo the after-dinner entertainment. On that night, 2 October 1611, Galileo staged a debate with a philosophy professor from Pisa, arguing on the subject of floating bodies for the edification of all present. Galileo's explanation of

Lyncean Academy
coat of arms

what made ice and other objects float in water differed sharply from the Aristotelian logic being taught in the universities, and his adroit verbal decimation of any opponent made for spectator sport at the Tuscan court.

'Before answering the adversaries' arguments,' a contemporary observer reported of Galileo's debating style, 'he amplified and reinforced them with apparently very powerful evidence which then made his adversaries look more ridiculous when he eventually destroyed their positions.'

The prevailing wisdom about bodies in water held that ice was heavier than water, but that broad, flat-bottomed pieces of ice floated anyway because of their shape, which failed to pierce the fluid surface. Galileo knew ice to be less dense than water, and therefore lighter, so that it always floated, regardless of its shape. He could show this by submerging a piece of ice and then releasing it under water to let it pop back up to the surface. Now, if shape were all that kept ice from sinking, then shape should also prohibit its upward motion through water – and all the more so if ice truly outweighed water.

Invited to join the discussion on floating bodies, Cardinal Barberini enthusiastically took Galileo's side. Later, he told Galileo in a letter: 'I pray the Lord God to preserve you, because men of great value like you deserve to live a long time to the benefit of the public.'

Cardinal Barberini had come to Florence to visit two of his nieces – both nuns, who lived at a local convent. This coincidence may have suggested a course to Galileo concerning his own two daughters, though the thought of placing them in a convent could have occurred to him naturally enough. Not only had his two sisters been schooled and sheltered in convents, but such institutions proliferated all around him. In Galileo's time, in addition to nearly thirty thousand males of all ages and more than thirty-six thousand females living in the city of Florence, a separately tallied population of 'religious' – one thousand men and four thousand women – dwelled in twenty-seven local monasteries and fifty-three convents. The pealing of bells from atop these cloistered residences reverberated through the air, day and night, as constant a note in the din of life as birdsong or conversation. Fully 50 per cent of the daughters of Florentine patrician families spent at least part of their lives within convent walls.

Galileo's sisters had eventually left the convent for holy matrimony, but he foresaw no such future for his daughters because of the conditions of their birth. At their present ages, eleven and ten, they were too young to take religious vows, yet they might well enter a convent before the canonical age of sixteen in any case, and bide the intervening years in a safer environment than he could provide for them, considering the plights of the women in his family: Madonna Giulia, always argumentative, had grown more difficult as she grew older, while his sisters were both burdened with their own young children and frequent pregnancies.

Galileo's poor health perhaps rushed his judgment on the matter, since he again took seriously ill within days of the court dispute over floating bodies and did not recover for several months. His illness forced him to flee the city for his private sanatorium at the Villa delle Selve, the country home of a generous good friend. From his bed in the hills, at the grand duke's behest, Galileo began putting his thoughts on floating bodies into a book-length treatment, to be called *Discourse on Bodies That Stay Atop Water or Move Within It*.

While at work on this project, he received a disturbing letter from an artist acquaintance of his in Rome: 'I have been told by a friend of mine, a priest who is very fond of you,' the painter Ludovico Cardi da Cigoli warned Galileo, 'that a certain crowd of ill-disposed men envious of your virtue and merits met at the house of the archbishop there [in Florence] and put their heads together in a mad quest for any means by which they could damage you, either with regard to the motion of the earth or otherwise. One of them wished to have a preacher state from the pulpit that you were asserting outlandish things. The priest, having perceived the animosity against you, replied as a good Christian and a religious man ought to do. Now I write this to you so that your eyes will be open to such envy and malice on the part of that sort of evildoers.'

These gathering storms may have confirmed Galileo's decision to cloister his daughters in the protective environment of a convent, for during the same period he wrote the letters that set the placement process in motion.

He insisted the girls stay together, despite the frowning of the Florentine Sacred Congregation of Bishops and Regulars on the question of admitting two siblings into the same convent. Although Galileo did not set down his reasons for his wish, he may well have seen Livia already displaying the morbid tendency to melancholy and withdrawal that would shade her adult personality. Without her sunny elder sister to counteract those dark moods, what would become of her? No other Italian city, Galileo learned, opposed the entry of natural sisters into the same monastery, but he would not send the girls to another city. He preferred to keep them close by, even if that meant seeking special dispensations.

'In answer to your letter concerning your daughters' claustration,' Francesco Maria Cardinal del Monte wrote to Galileo in December of 1611,

> I had fully understood that you did not wish them to take the veil immediately, but that you wished them to be received on the understanding that they were to assume the religious habit as soon as they had reached the canonical age. But, as I have written to you before, even this is not allowed, for many reasons: in particular, that it might give rise to the exercise of undue influence by those who wished the young persons to take the veil for reasons of their own. This rule is never broken, and never will be, by the Sacred Congregation. When they have reached the canonical age, they may be accepted with the ordinary dowry, unless the sisterhood already has the prescribed number; if such be the case, it will be necessary to double the dowry. Vacancies may not be

filled up by anticipation under severe penalties, that of deprivation for the Abbess in particular, as you may see in a Decretal of Pope Clement of the year 1604.

It could never be done, but it happened all the time, as Galileo was well aware. If Cardinal del Monte, who had finessed Galileo's first teaching appointment at Pisa, proved unwilling or unable to get the two girls into one convent before either of them turned sixteen, then some other contact might yet intervene.

As he neared completion of his treatise on floating bodies, Galileo penned an explanation to the grand duke and the general public as to why his new book concerned bodies in water, instead of continuing the great chain of astronomical discoveries trumpeted in *The Starry Messenger*. Lest anyone think he had dropped his celestial observations or pursued them too slowly, he could account for his time. 'A delay has been caused not merely by the discoveries of three-bodied Saturn and those changes of shape by Venus resembling the moon's, along with consequences that follow thereon,' Galileo wrote in the introduction, 'but also by the investigation of the times of revolution around Jupiter of each of the four Medicean planets, which I managed in April of the past year, 1611, while I was at Rome . . . I add to these things the observation of some dark spots that are seen in the Sun's body . . . Continued observations have finally assured me that such spots are . . . carried around by rotation of the Sun itself, which completes its period in about a lunar month – a great event, and even greater for its consequences.'

Thus *Bodies in Water* not only challenged Aristotelian physics on the behaviour of submerged or floating objects but also defaced the perfect body of the Sun. Galileo further flouted academic tradition by writing *Bodies in Water* in Italian, instead of the Latin lingua franca that enabled the European community of scholars to communicate among themselves.

'I wrote in the colloquial tongue because I must have everyone able to read it,' Galileo explained – meaning the shipwrights he admired at the Venetian Arsenale, the glassblowers of Murano, the

lens grinders, the instrument makers, and all the curious com-
patriots who attended his public lectures. 'I am induced to do this
by seeing how young men are sent through the universities at
random to be made physicians, philosophers, and so on; thus
many of them are committed to professions for which they are
unsuited, while other men who would be fitted for these are taken
up by family cares and other occupations remote from literature
. . . Now I want them to see that just as Nature has given to them,
as well as to philosophers, eyes with which to see her works, so
she has also given them brains capable of penetrating and understanding them.'

Galileo's behaviour enraged and insulted his fellow philosophers – especially those, like Ludovico delle Colombe of the
Florentine Academy, who had tussled with him in public and lost.
Colombe declared himself 'anti-Galileo' in response to Galileo's
anti-Aristotelian stance. Supporters of Galileo, in turn, took up the
title 'Galileists' and further deflated Colombe's flimsy philosophy
by playing derisively on his name. Since *colombe* means 'doves' in
Italian, they dubbed Galileo's critics 'the pigeon league'.

[V]

In the very face of the Sun

IT IS DIFFICULT TODAY – from a vantage point of insignificance on this small planet of an ordinary star set along a spiral arm of one galaxy among billions in an infinite cosmos – to see the Earth as the centre of the universe. Yet that is where Galileo found it.

The cosmology of the sixteenth and seventeenth centuries, founded on the fourth-century-BC teachings of Aristotle and refined by the second-century Greek astronomer Claudius Ptolemy, made Earth the immobile hub. Around it, the Sun, the Moon, the five planets and all the stars spun eternally, carried in perfectly circular paths by the motions of nested crystalline celestial spheres. This heavenly machinery, like the gearwork of a great clock, turned day to night and back to day again.

In 1543, however, the Polish cleric Nicolaus Copernicus flung the Earth from its central position into orbit about the Sun, in his book *On the Revolutions of the Heavenly Spheres*, or *De revolutionibus*, as it is usually called. By imagining the Earth to turn on its own axis once a day, and travel around the Sun once a year, Copernicus rationalised the motions of the heavens. He saved the enormous Sun the trouble of traipsing all the way around the smaller Earth from morning till evening. Likewise the vast distant realm of the stars could now lie still, instead of having to wheel overhead even more rapidly than the Sun every single day. Copernicus also called the planets to order, relieving those bodies of the need to co-ordinate their relatively slow motion towards the east over long periods of time (Jupiter takes twelve years to traverse the twelve constellations of the zodiac, Saturn thirty) with their speedy west-wards day trips around the Earth. Copernicus could even explain the way Mars, for example, occasionally reversed its course, drift-ing *backwards* (westwards) against the background of the stars for months at a time, as the logical consequence of heliocentrism: the Earth occupied an inside track among the paths of the planets – third from the Sun, as opposed to Mars's fourth position – and could thus overtake the slower, more distant Mars every couple of years.

Copernicus, who studied astronomy and mathematics at the University of Cracow, medicine for a while in Padua and canon law in Bologna and Ferrara, devoted most of his life to cosmology, thanks to nepotism. When he returned to Poland from his studies in Italy at the age of thirty, his uncle, a bishop, helped secure Copernicus a lifetime appointment as a canon at the cathedral of Frombork. Serving forty years in that 'most remote corner of the Earth', with manageable duties and a comfortable pension, Coper-nicus created an alternative universe.

'For a long time I reflected on the confusion in the astronomical traditions concerning the derivation of the motion of the spheres of the Universe,' Copernicus wrote in Frombork. 'I began to be annoyed that the philosophers had discovered no sure scheme for the movements of the machinery of the world, created for our sake by the best and most systematic Artist of all. Therefore, I began to consider the mobility of the Earth and even though the idea seemed absurd, nevertheless I knew that others before me had been granted the freedom to imagine any circles whatsoever for explaining the heavenly phenomena.'

Although he made numerous naked-eye observations of the positions of the planets, most of Copernicus's lonely work involved reading, thinking and mathematical calculations. He proffered no supporting evidence of any kind. And nowhere, alas, did he record the train of thought that led him to his revolutionary hypothesis.

An anonymous introductory note to Copernicus's book dismissed the whole conceit as merely an aid to computation. The complex business of determining the orbital periods of the planets, including the Sun and Moon, figured crucially in establishing the length of the year and the date of Easter. Copernicus himself, writing in the languages of Latin and mathematics for a scholarly audience, never attempted to convince the general public that the universe was actually constructed with the Sun at the centre. And who would have believed him if he had? The fact that the Earth remained motionless was a truism, obvious to any sentient individual. If the Earth rotated and revolved, then a ball tossed into the air would not fall right back into one's hands but land hundreds of feet away, birds in flight might lose the way to their nests, and all humanity suffer dizzy spells from the daily spinning of the global carousel at one thousand miles per hour.*

'The scorn which I had to fear', Copernicus remarked in *De revolutionibus*, 'on account of the newness and absurdity of my opinion almost drove me to abandon a work already undertaken.' Continuous calculation and checking delayed publication of his manuscript for decades, until he lay literally on his deathbed. Expiring at the age of seventy, immediately after the first printing of his book in 1543, Copernicus avoided any brush with derision.

When Galileo ascended the wooden steps of his teaching plat-

---

* This is the actual speed of the Earth's rotation at the Equator. Its speed of revolution about the Sun exceeds seventy thousand miles per hour.

form at Padua to lecture on planetary astronomy, beginning in 1592, he taught the Earth-centred view, as it had been preserved from antiquity. Galileo knew of Copernicus's challenge to both Aristotle and Ptolemy, and he may have casually mentioned this alternative idea to his students, too. Heliocentrism, however, played no part in his formal curriculum, which was primarily concerned with teaching medical students how to cast horoscopes. Nevertheless, Galileo gradually convinced himself that the Copernican system not only looked neater on paper but was likely to hold true in fact. In a 1597 letter he wrote to a former colleague at Pisa, Galileo assessed the system of Copernicus as 'much more probable than that other view of Aristotle and Ptolemy'. He expressed the same faith in Copernicus in a letter he wrote to Kepler later that year, regretting how 'our teacher Copernicus, who though he will be of immortal fame to some, is yet by an infinite number (for such is the multitude of fools) laughed at and rejected'. Since the Copernican system remained just as absurd to popular opinion fifty years following its author's demise, Galileo long maintained his public silence on the subject.

In 1604, five years prior to Galileo's development of the telescope, the world beheld a never-before-seen star in the heavens. It was called 'nova' for its newness.* It flared up near the constellation Sagittarius in October and stayed so prominent through November that Galileo had time to deliver three public lectures about the newcomer before it faded from bright view. The nova challenged the law of immutability in the heavens, a cherished tenet of the Aristotelian world order. Earthly matter, according to ancient Greek philosophy, contained four base elements – earth, water, air, fire – that underwent constant change, while the heavens, as Aristotle described them, consisted entirely of a fifth element – the quintessence, or aether – that remained incorruptible. It was thus impossible for a new star suddenly to materialise.

---

* Modern astronomers define a nova as the sudden dramatic brightening of an otherwise unseen star. What Galileo saw in 1604 would today be termed a 'supernova' – the fireball explosion of a dying star.

The nova, the Aristotelians argued, must inhabit the sublunar sphere between the Earth and the Moon, where change was permissible. But Galileo could see by comparing his nightly observations with those of other stargazers in distant lands that the new star lay far out, beyond the Moon, beyond the planets, among the domain of the old stars.

In his playful, provocative way, Galileo presented the nova controversy to the public in a dialogue between a pair of peasants speaking Paduan dialect, which he published under the pen name Alimberto Mauri. Call the new star 'quintessence', his gruff hero concluded, or call it 'polenta'! Careful observers could measure its distance just the same.

Having thus impugned the immutability of the heavens, Galileo further attacked the defensive Aristotelian philosophers by turning the telescope on their territory in 1609. His telescopic discoveries transformed the nature of the Copernican question from an intellectual engagement into a debate that might be decided on the basis of evidence. The roughness of the Moon, for example, showed that some of the features of Earth repeated themselves in the heavens. The motions of the Medicean stars demonstrated that satellites could orbit bodies other than the Earth. The phases of Venus argued that at least one planet must travel around the Sun. And the dark spots discovered on the Sun sullied the perfection of yet another heavenly sphere. 'In that part of the sky which deserves to be considered the most pure and serene of all – I mean in the very face of the sun,' Galileo reported, 'these innumerable multitudes of dense, obscure and foggy materials are discovered to be produced and dissolved continually in brief periods.'

Galileo rued the stubbornness of philosophers who clung to Aristotle's views despite the new perspective provided by the telescope. He swore that if Aristotle himself were brought back to life and shown the sights now seen, the great philosopher would quickly alter his opinion, as he had always honoured the evidence of his senses. Galileo chided the followers of Aristotle for being too timid to stray from their master's texts: 'They wish never to raise their eyes from those pages – as if this great book of the universe had been written to be read by nobody but Aristotle, and his eyes had been destined to see for all posterity.'

Several of Galileo's Aristotelian opponents sputtered that the sunspots must be a new fleet of 'stars' circling the Sun the way the Medicean stars orbited Jupiter. Even professors who had vociferously rejected the moons of Jupiter, damning them as demonic visions spawned by the distorting lenses of Galileo's telescope, now turned to embrace them as the Sun's last hope for maintaining its steady stateliness.

One of the first scientists to see sunspots, Galileo gathered important correspondents among foreign astronomers seeking to compare observations and interpretations with him. In January of 1612, while still convalescing at the Villa delle Selve outside Florence, Galileo heard much about sunspots from a German gentleman and amateur scientist named Marcus Welser. 'Most Illustrious and Excellent Sir,' Welser hailed Galileo,

> Already the minds of men are assailing the heavens, and gain strength with every acquisition. You have led in scaling the walls, and have brought back the awarded crown. Now others follow your lead with the greater courage, knowing that once you have broken the ice for them it would indeed be base not to press so happy and honourable an undertaking. See, then, what has arrived from a friend of mine; and if it does not come to you as anything really new, as I suppose, nevertheless I hope you will be pleased to see that on this side of the mountains also men are not lacking who travel in your footsteps. With respect to these solar spots, please do me the favour of telling me frankly your opinion – whether you judge them to be made of starry matter or not; where you believe them to be situated, and what their motion is.

Enclosed Galileo found several essays by Welser's 'friend', an anonymous astronomer (later revealed as Father Christopher Scheiner, Jesuit professor at the University of Ingolstadt), who tried to explain the new phenomenon according to the old philosophy, protecting his identity behind the pseudonym 'Apelles'.

Galileo took nearly four months to formulate his reply, constrained at first by his illness ('a long indisposition,' he called it, 'or I should say a series of long indispositions preventing all exercises and occupations on my part'), and even further by the calumny of his enemies, not to mention the mysterious nature of the spots themselves.

'The difficulty of this matter,' Galileo finally conceded to Welser, 'combined with my inability to make many continued observations, has kept (and still keeps) my judgment in suspense. And I, indeed, must be more cautious and circumspect than most other people in pronouncing upon anything new. As Your Excellency well knows, certain recent discoveries that depart from common and popular opinions have been noisily denied and impugned, obliging me to hide in silence every new idea of mine until I have more than proved it.' Nevertheless, Galileo expounded on the essence and substance of sunspots for many pages, initiating an ongoing correspondence with Welser – and through him 'the masked Apelles' – that sounded the full thunder of the new debate. Indeed, Galileo's letters on sunspots speak almost as much about the system of the world as they do about the solar spots.

'With absolute necessity we shall conclude,' Galileo wrote early in the first of his three letters to Welser, 'in agreement with the theories of the Pythagoreans and of Copernicus, that Venus revolves about the Sun just as do all the other planets ... No longer need we employ arguments that allow any answer, however feeble, from persons whose philosophy is badly upset by this new arrangement of the universe.'

Apelles upheld the idea that the dark spots must be many small stars circling the Sun. Galileo saw nothing starlike about them. To his mind, they more closely resembled clouds: 'Sunspots are generated and decay in longer and shorter periods; some condense and others greatly expand from day to day; they change their shapes, and some of these are most irregular; here their obscurity is greater and there less. They must be simply enormous in bulk, being either on the Sun or very close to it. By their uneven opacity they are capable of impeding the sunlight in differing degrees; and sometimes many spots are produced, sometimes few, sometimes none at all.'

But he quickly added: 'I do not assert on this account that the spots are clouds of the same material as ours, or aqueous vapours raised from the Earth and attracted by the Sun. I merely say that we have no knowledge of anything that more closely resembles them. Let them be vapours or exhalations then, or clouds, or fumes sent out from the Sun's globe or attracted there from other places; I do not decide on this – and they may be any of a thousand other things not perceived by us.' (He could never have imagined, despite his long-standing interest in magnets, that the spots marked the sites of the Sun's most potent magnetic fields.)

'If I may give my own opinion to a friend and patron,' Galileo continued, 'I shall say that the solar spots are produced and dissolve upon the surface of the Sun and are contiguous to it, while the Sun, rotating upon its axis in about one lunar month, carries them along, perhaps bringing back some of those that are of longer duration than a month, but so changed in shape and pattern that it is not easy for us to recognise them.'

In closing this first letter, Galileo begged Welser's indulgence:

And forgive me my indecision, because of the novelty and difficulty of the subject, in which various thoughts have passed through my mind and met now with assent and again with rejection, leaving me abashed and perplexed, for I do not like to open my mouth without declaring anything whatever. Nevertheless, I shall not abandon the task in despair. Indeed, I hope that this new thing will turn out to be of admirable service in tuning for me some reed in this great discordant organ of our philosophy – an instrument on which I think I see many organists wearing themselves out trying vainly to get the whole thing into perfect harmony. Vainly, because they leave (or rather preserve) three or four of the principal reeds in discord, making it quite impossible for the others to respond in perfect tune.

These off-key reeds that Galileo decried sounded several flat notes, including the immutability of the heavens, the farrago of the celestial spheres and the immobility of the Earth.

Welser wrote back gratefully to say, 'You have paid a high rate of interest for the favour of a little time, sending me so copious and diffuse a treatise in reply to a few lines.' The thrill of witnessing new philosophy grow up around the astronomical anomaly of the sunspots made Welser want to share Galileo's letter with a larger audience than just the alleged Apelles, who didn't even read Italian and had to wait months for a suitable translation to be made. Welser thought perhaps Prince Cesi of the Lyncean Academy, with whom he also corresponded, should publish the sunspot report as part of an ongoing series. 'It would be a public benefit for these little treatises concerning new discoveries to come out one by one,' Welser opined, 'keeping things fresh in everyone's mind and inspiring others to apply their talents more to such things; for it is impossible that so great a framework should be sustained upon the shoulders of one man, however strong.'

Prince Cesi liked the idea so well that he not only initiated preparations for printing but also inducted Welser into the academy. Soon Welser and Galileo were both signing themselves proudly as 'Lyncean' in their letters and politely commiserating with each other on their physical complaints. When Cesi published Welser's four relatively short notes together with Galileo's three very long replies as *History and Demonstrations Concerning Sunspots and Their Phenomena* in Rome in the spring of 1613, he retained all the chitchat about Welser's gout and Galileo's miscellaneous infirmities.

'I have read [your letter], or rather devoured it, with a pleasure equal to the appetite and longing I had for it,' Welser wrote to Galileo on 1 June 1612. 'Let me assure you that it has served to alleviate for me a long and painful illness that has been causing me extreme discomfort in the left thigh. For this the physicians have not yet found any effective remedy; indeed, the doctor in charge has told me in very plain words that the first men of his profession have written of this disease that "some cases are cured, but others

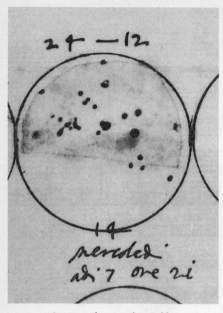

Sunspot drawing by Galileo

are incurable". One must therefore submit to the fatherly disposition of God's providence; "Thou art the Lord, do what is good in Thy sight."'

Poor Welser would be dead within two years, escaping the pain of his disease through suicide, but meanwhile he worried how Cesi could accomplish the printing of the many meticulous drawings, which Galileo appended to his letters, of sunspots ingeniously observed. Galileo rendered these near-photographic records by letting the telescope image of the Sun fall on a piece of white paper instead of on his retina. There he faithfully traced them – and later retraced them, reorienting the telescope's inverted vision – to avoid any damage to his eyes.

More than a month's worth of excellent engravings embellished the finished book, tracking the Sun day by day from the 1st of June until mid-July of 1612. The ideas espoused in the book, however, exacerbated the existing tensions between Galileo and his avowed opponents. Book discussions attracted additional new opponents among people who had not even read the text. And, since Copernicus had died in silence in another country years before, Galileo began to be credited – or rather blamed – for having fathered the Sun-centred universe.

Although the 'pigeon league' attacks on *Bodies in Water* had held up the books of Aristotle to oppose Galileo, critics of the *Sunspot Letters* now appealed to the even higher authority of the Bible.

[VI]

Observant executrix of God's commands

THE REORGANISATION OF THE heavens according to Copernicus struck some individuals as suspiciously heretical.

'That opinion of Ipernicus, or whatever his name is,' an elderly Dominican father wagged in Florence in November of 1612, 'appears to be against Holy Scripture.' Neither Copernicus nor Galileo, however, both Catholic believers, intended any such criticism of the Bible or attack against the Church. Copernicus had in fact dedicated *De revolutionibus* to Pope Paul III (the pontiff who excommunicated Henry VIII and established the Roman Inquisition). Galileo, in the course of writing the *Sunspot Letters*, had sought the expert opinion of Carlo Cardinal Conti on the subject of change in the heavens. Cardinal Conti had assured him that the Bible did not support Aristotle's doctrine of immutability; in fact, he said, Scripture seemed to argue against it.

None of his experience parrying angry attacks from academics prepared Galileo for the intimations of heresy – a crime he considered 'more abhorrent than death itself' – that now swirled around him. Given these circumstances, he must have felt relieved in October 1613, when Ottavio Cardinal Bandini, another prelate of his acquaintance, finally secured the dispensation of age for Galileo's daughters. The immediate admission of both thirteen-year-old Virginia and twelve-year-old Livia into the nearby Convent of San Matteo in Arcetri was apparently facilitated by the coincidence that the mother abbess, Suor Ludovica Vinta, was sister to a Florentine senator who had served as secretary of state under Grand Duke Ferdinando. No sooner were the girls secured within the enclosure walls than the pitch of the Copernican controversy escalated.

In November, Galileo's best and most beloved student, the Benedictine monk Benedetto Castelli, who had followed him from Padua, left Florence to take over Galileo's old chair of mathematics at the University of Pisa. Castelli not only had devised the safe method of observing the Sun on paper used by Galileo to such good effect, but had actually drawn the numerous sunspot diagrams published in Galileo's book. Galileo had further relied on Castelli to answer all four published attacks on *Bodies in Water*. Newly arrived at Pisa, Castelli was warned by the university overseer never to teach or even discuss the motion of the Earth. The monk agreed to these terms, naturally, pointing out that his mentor, Galileo, had followed the same course throughout two-plus decades of lecturing at both Pisa and Padua. Within weeks, however, Castelli found himself specifically questioned on the matter of Copernicus in a private but most influential setting, when the Medici family and full entourage arrived in Pisa for their annual winter visit. Holding court for the season at their Pisan palace, Their Serene Highnesses Cosimo II, Archduchess Maria Maddalena and Grand Duchess Mother Madama Cristina filled the seats around their table three times a day with interesting conversationalists who could inform them on a variety of subjects.

Benedetto Castelli

'Thursday morning I was dining with our Patrons,' Castelli wrote to Galileo on Saturday 14 December, 'and when asked about the university by the Grand Duke I gave him a complete account of everything, with which he showed himself much pleased. He asked me if I had a telescope; saying yes, I began to tell about an observation of the Medicean planets I had made just the night before. Madama Cristina wanted to know their position, whereupon the talk turned to the necessity of their being real objects and not illusions of the telescope.'

Instead of receding from court life following the death of her husband, Ferdinando I, in 1609, the influential grand duchess Cristina had changed her dress to black and donned a widow's cap with a voluminous black veil in place of her ducal crown. She had held fast to her rank of grand duchess, leaving her daughter-in-law – Cosimo's wife, Maria Maddalena – to be content with the 'archduchess' title that had come with her from Austria.

On this particular December morning, Madama Cristina found Castelli's talk of planets disturbing – despite their ties to the House of Medici. Notwithstanding her fondness for Galileo, who had tutored her son, and over and above her respect for Castelli's monastic robes, she much preferred the conversation of another breakfast guest from the university faculty, the Platonic philosopher Doctor Cosimo Boscaglia.

'After many things, all of which passed with decorum,' Castelli's letter continued, 'breakfast was over. I left, but I had hardly come out of the palace when I was overtaken by the porter of Madama Cristina, who had recalled me. But before I tell you what followed, you must first know that while we were at table Doctor Boscaglia had had Madama's ear for a while, and while conceding as real all the things you have discovered in the sky, he said that only the motion of the Earth had in it something of the incredible, and could not occur, especially because the Holy Scripture was obviously contrary to that view.'

Friends of the court all knew Madama Cristina to be a devout Catholic who lent her ear most frequently to her confessor, other priests, cardinals and of course the pope, even when His Holiness's opinions ran counter to the best interests of the Medici dynasty or the Tuscan government. She had read her Bible and could quote from the Book of Joshua – wherein the Sun is ordered to stand still, presumably because it had been moving – as well as the Psalms:

> O Lord my God, Thou art great indeed . . . Thou fixed the Earth upon its foundation, not to be moved for ever.
>
> PSALM 104: 1,5

'Now, getting back to my story,' Castelli went on,

> I entered into the chambers of her Highness, and there I found the Grand Duke, Madama Cristina and the Archduchess, Don Antonio [de' Medici], Don Paolo Giordano [Orsini], and Doctor Boscaglia. Madama began, after some questions about myself, to argue the Holy Scripture against me. Thereupon, after having made suitable disclaimers, I commenced to play the theologian with such assurance and dignity that it would have done you good to hear me. Don Antonio assisted me, giving me such heart that instead of being dismayed by the majesty of their Highnesses I carried things off like a paladin. I quite won over the Grand Duke and his

Archduchess, while Don Paolo came to my assistance with a very apt quotation from the Scripture. Only Madama remained against me, but from her manner I judged that she did this only to hear my replies. Professor Boscaglia said never a word.

The troubling news of Madama Cristina's displeasure inspired an immediate response from Galileo. Even more than he regretted her opposition, he dreaded the drawing of battle lines between science and Scripture. Personally, he saw no conflict between the two. In the long letter he wrote back to Castelli on 21 December 1613, he probed the relationship of discovered truth in Nature to revealed truth in the Bible.

'As to the first general question of Madama Cristina, it seems to me that it was most prudently propounded to you by her, and conceded and established by you, that Holy Scripture cannot err and the decrees therein contained are absolutely true and inviolable. I should only have added that, though Scripture cannot err, its expounders and interpreters are liable to err in many ways . . . when they would base themselves always on the literal meaning of the words. For in this wise not only many contradictions would be apparent, but even grave heresies and blasphemies, since then it would be necessary to give God hands and feet and eyes, and human and bodily emotions such as anger, regret, hatred and sometimes forgetfulness of things past, and ignorance of the future.'

These literary devices had been inserted into the Bible for the sake of the masses, Galileo insisted, to aid their understanding of matters pertaining to their salvation. In the same way, biblical language had also simplified

Grand Duchess Cristina of Lorraine

certain physical effects in Nature, to conform to common experience. 'Holy Scripture and Nature', Galileo declared, 'are both emanations from the divine word: the former dictated by the Holy Spirit, the latter the observant executrix of God's commands.'

Thus no truth discovered in Nature could contradict the deep truth of Holy Writ. Even Madama Cristina's objection regarding the Book of Joshua could be put to rest in terms of the Sun-centred universe; indeed, Copernicus made more sense of the passage than either Aristotle or Ptolemy, as Galileo spent almost half of this letter explaining.

> On this day, when the Lord delivered up the Amorrites to the Israelites, Joshua prayed to the Lord, and said in the presence of Israel: Stand still, O sun at Gabhaon, O moon, in the valley of Aialon! And the sun stood still, and the moon stayed, while the nation took vengeance on its foe. Is this not recorded in the Book of Hashar? The sun halted in the middle of the sky; not for a whole day did it resume its swift course. Never before or since was there a day like this, when the Lord obeyed the voice of a man; for the Lord fought for Israel.
>
> JOSH. 10: 12–14

The Ptolemaic system granted the Sun two motions. One of these, a slow annual progression from west to east, belonged strictly to the Sun itself. The other, more apparent, motion of the Sun across the sky over the course of the day – most probably the motion Joshua had sought to halt – actually belonged to the Ptolemaic Primum Mobile, the sphere of the highest sky, which spun all the other spheres containing Sun, Moon, planets and stars around the Earth every twenty-four hours. God's stopping only the Sun would not have achieved Joshua's desire. On the contrary, it would have made night arrive about four minutes early.

As Copernicus viewed the sky, however, the passage of day to night resulted from the turning of the Earth. Galileo agreed with Copernicus that the Earth somehow drew this motion from the Sun. Galileo had further observed the Sun to have its own monthly rotation, which he discovered during his studies of sun-spots. Just as the light of the Sun illuminated all the planets, so too its motion energised them to pursue their orbits. Therefore, if God had stopped the Sun's rotation, the Earth would have stopped, too, and the day stretched out to accommodate Joshua's needs.

Later Galileo would point out that when the Sun stood still in the biblical account, it did so 'in the middle of the sky' – precisely where the Copernican system placed it. This reference to location could not be taken to mean the Sun had been standing in the high noontime position, for then Joshua would have found time enough to fight his battle without praying for a miracle to prolong the day.

Despite the strength of his argument, Galileo personally wished to abandon all such astronomical interpretations, on the grounds that the Bible spoke to a more important purpose. As he had once heard the late Vatican librarian Cesare Cardinal Baronio remark, the Bible was a book about how one goes to Heaven – not how Heaven goes.

'I believe that the intention of Holy Writ was to persuade men of the truths necessary for salvation,' Galileo continued his letter to Castelli, 'such as neither science nor any other means could render credible, but only the voice of the Holy Spirit. But I do not think it necessary to believe that the same God who gave us our senses, our speech, our intellect, would have put aside the use of these, to teach us instead such things as with their help we could find out for ourselves, particularly in the case of these sciences of which there is not the smallest mention in the Scriptures; and, above all, in astronomy, of which so little notice is taken that the names of none of the planets are mentioned. Surely if the intention of the sacred scribes had been to teach the people astronomy, they would not have passed over the subject so completely.'

Castelli shared this exquisite exposition with friends and col-leagues, who hand-copied it and forwarded it numerous times. Galileo now returned to predicting the positions of the Medicean satellites and to penning responses to various published attacks against his own published works. When his health faltered in

March, Castelli, who had been begging Galileo to take better care of himself, stepped in to help him.

In the early days of summer, at the Convent of San Matteo in Arcetri, Virginia and Livia began to wear the dark-brown religious habits of the Franciscan orders. Although both children were still too young to take their vows, mother abbess Suor Ludovica Vinta told the ailing Galileo that she desired to see them appropriately outfitted before she relinquished her elected office.

> *Young girls received into the monastery before the age required by law shall have their hair cut off round and, their secular dress being laid aside, shall be clothed in religious garb as it shall seem fitting to the Abbess. But when they have reached the age required by law, they shall make their profession clothed after the manner of the others.*
>
> THE RULE OF SAINT CLARE, chapter II

Meanwhile Galileo's letter to Castelli continued to circulate, travelling from hand to hand, and eventually falling into the wrong hands. On 21 December 1614, exactly one year to the day after Galileo wrote the letter, he found himself denounced from the pulpit of the Church of Santa Maria Novella, right in the city of Florence, by Tommaso Caccini, a hotheaded young Dominican priest with ties to the 'pigeon league'.

> *Men of Galilee, why do ye stand looking up to heaven?*
>
> ACTS I: II

Beginning his sermon with this barb, Caccini moved quickly to the biblical text for that Advent Sunday, which happened to come from the Book of Joshua, and included the 'Stand still, O sun' command that had sparked Madama Cristina's original complaint. Caccini wound up branding Galileo, Galileo's followers, and all mathematicians in general 'practitioners of diabolical arts . . . enemies of true religion'.

The vitriol of the language brought Caccini a reprimand, and

Galileo a written apology from the preacher's Dominican superior. But soon another Florentine Dominican, Niccolò Lorini, submitted a copy of Galileo's now widely read letter to Castelli to an inquisitor general in Rome. Hearing this news, Galileo feared that crucial passages might have been altered (as indeed proved to be the case) either by mistakes in copying or through malevolent distortion. He sent a true copy to his friend at the Vatican, Piero Dini, who in turn copied it repeatedly for various cardinals who might help clear Galileo's name.

Through the spring and summer of 1615, Galileo sustained yet another long bout of incapacitating illness – aggravated, perhaps, by his recognition of the forces arrayed against him. Indeed he saw himself the focus of a conspiracy. While bedridden, he recast his informal letter to Castelli into a much longer, more referenced treatise addressed to Madama Cristina herself. (Though no printer dared publish the *Letter to the Grand Duchess Cristina* until 1636, in Strasbourg, manuscript copies enjoyed a wide Italian readership.) 'Some years ago,' he began,

> as Your Serene Highness well knows, I discovered in the heavens many things that had not been seen before our own age. The novelty of these things, as well as some consequences which followed from them in contradiction to the physical notions commonly held among academic philosophers, stirred up against me no small number of professors – as if I had placed these things in the sky with my own hands in order to upset Nature and overturn the sciences. They seemed to forget that the increase of known truths stimulates the investigation, establishment and growth of the arts; not their diminution or destruction.
>
> Showing a greater fondness for their own opinions than for truth, they . . . hurled various charges and published numerous writings filled with vain arguments, and they made the grave mistake of sprinkling these with passages taken from places in the Bible which they had failed to understand properly, and which were ill suited to their purposes.

Even though Galileo directed these comments to Madama Cristina, he refrained from accusing her of the same injustices, which she had committed without malice. He reserved his venom for those others who used biblical passages they could not comprehend to condemn the worthy theory of Copernicus, which they had not read. He backed his position by quoting Saint Augustine, who advised moderation in piety and caution in judgment on complex issues, so as to avoid condemning hypotheses 'that truth hereafter may reveal to be not contrary in any way to the sacred books of either the Old or the New Testament'. In the margins of his fifty-page letter, Galileo footnoted all the theological works he had consulted to construct his thesis concerning the use of biblical quotations in matters of science – allowing the likes of Saint Augustine, Tertullian, Saint Jerome, Saint Thomas Aquinas, Dionysius the Areopagite and Saint Ambrose to defend him against enemies who sought 'to destroy me and everything mine by any means they can think of'.

Galileo felt he understood the motivation of his detractors: 'Possibly because they are disturbed by the known truth of other propositions of mine which differ from those commonly held, and therefore mistrusting their defence so long as they confine themselves to the field of philosophy, these men have resolved to fabricate a shield for their fallacies out of the mantle of pretended religion and the authority of the Bible.'

The Holy Fathers of the Church of course occupied a separate category. Yet several of these, Galileo complained, usurped scriptural authority to pronounce judgments in physical disputes, while ignoring any evidence of science to the contrary:

> Let us grant then that theology is conversant with the loftiest divine contemplation, and occupies the regal throne among sciences by dignity. But acquiring the highest authority in this way, if she does not descend to the lower and humbler speculations of the subordinate sciences and has no regard for them because they are not concerned with blessedness, then her professors should not arrogate to themselves the

authority to decide on controversies in professions which they have neither studied nor practised. Why, this would be as if an absolute despot, being neither a physician nor an architect but knowing himself free to command, should undertake to administer medicines and erect buildings according to his whim – at grave peril of his poor patients' lives, and the speedy collapse of his edifices.

Galileo took pains to establish the antiquity of the Sun-centred universe, which dated all the way back to Pythagoras in the sixth century BC, was later upheld by Plato in his old age, and also adopted by Aristarchus of Samos, as reported by Archimedes in the *Sand-reckoner*, before being codified by the Catholic canon Copernicus in 1543. Galileo had good reason to suspect that this theory stood on the verge of suppression, and his *Letter to the Grand Duchess Cristina* argued passionately against such action:

To ban Copernicus now that his doctrine is daily reinforced by many new observations and by the learned applying themselves to the reading of his book, after this opinion has been allowed and tolerated for those many years during which it was less followed and less confirmed, would seem in my judgment to be a contravention of truth, and an attempt to hide and suppress her the more as she revealed herself the more clearly and plainly. Not to abolish and censure his whole book, but only to condemn as erroneous this particular proposition, would (if I am not mistaken) be a still greater detriment to the minds of men, since it would afford them occasion to see a proposition proved that it was heresy to believe. And to prohibit the whole science would be but to censure a hundred passages of Holy Scripture which teach us that the glory and greatness of Almighty God are marvellously discerned in all His works and divinely read in the open book of Heaven. For let no one believe that reading the lofty concepts written in that book leads to nothing further than the mere seeing of the splendour of the Sun and the stars and their rising and setting, which is as far as the eyes of brutes and of the vulgar can penetrate. Within its pages are couched mysteries so profound and concepts so sublime that the vigils, labours and studies of hundreds upon hundreds of the most acute minds have still not pierced them, even after continual investigations for thousands of years.

Having hereby framed his thoughts on paper, Galileo felt the gravity of the situation propelling him to Rome, where he intended to free his reputation of any whisper of heresy, and also to defend the burgeoning study of astronomy with new weapons of his own devising.

Grand Duke Cosimo gave Galileo permission to make the journey – over the objections of his Tuscan ambassador there, who judged Rome a dangerous place for the court philosopher 'to argue about the Moon'. Corridors leading to the Vatican and the Holy Office of the Inquisition already hummed with the controversy of his doctrines.

[V I I]

The malice
of my
persecutors

GALILEO ISSUED HIS CALL for a distinction between questions of science and articles of faith at an anxious moment in Church history.

Stunned by the Protestant Reformation fomented in Germany around 1517, the Roman Church struck a defensive posture throughout the sixteenth and seventeenth centuries called the Counter-Reformation. The Church hoped quickly to close the rift that had split Protestantism from Catholicism by convening an ecumenical council, but intrigues and obstacles of all sorts – including disputes over where to stage the event – postponed the meeting for many years, while the rift continued to widen. Finally

Pope Paul III (the same pontiff honoured in the dedication of Copernicus's book) convened bishops, cardinals and leaders of religious orders at Trent, where Italy bordered the Holy Roman Empire of the German nation. On and off over a period of eighteen years, from 1545 to 1563, the Council of Trent debated and voted and ultimately drafted a series of decrees.* These dictated how the clergy were to be educated, for example, and who was empowered to interpret Holy Scripture. Rejecting Martin Luther's insistence on the right to a personal reading of the Bible, the council declared in 1546 that 'no one, relying on his own judgment and distorting the Sacred Scriptures according to his own conceptions, shall dare to interpret them'.

After the council finally concluded the twenty-five sessions of its long-drawn-out deliberations, its decrees became Church doctrine through a series of papal bulls (so named after the *bulla*, or round lead seal, affixed to pronouncements from the pope himself). In 1564, the year Galileo was born, certain important points from the debates were formulated into a profession of faith, worded by the Council of Trent and solemnly sworn over the ensuing decades by untold numbers of Church officials and other Catholics:

> *I most firmly accept and embrace the Apostolic and ecclesiastical traditions and the other observances and constitutions of the Church. I also accept Sacred Scripture in the sense in which it has been held, and is held, by Holy Mother Church, to whom it belongs to judge the true sense and interpretation of the Sacred Scripture, nor will I accept or interpret it in any way other than in accordance with the unanimous agreement of the Fathers.*

Galileo's *Letter to the Grand Duchess Cristina* indirectly charged his opponents with violating this oath by bending the Bible to their purposes. His opponents, on the other hand, judged Galileo guilty

---

* Given the longevity of the council, its membership naturally changed considerably over the years, while ultimate approval of its decisions passed from Paul III to Julius III to Pius IV.

of the same crime. His only hope of winning the argument lay in producing proof positive for the Copernican system. Then, since no truth found in Nature could contradict the truth of Scripture, everyone would realise that the Fathers' judgment about the placement of the heavenly bodies had been hasty, and required reinterpretation in the light of scientific discovery.

December 1615 thus brought Galileo to Rome brandishing new support for Copernicus – derived from observations of the *Earth*, not the heavens. The tidal motions of the great oceans, Galileo believed, bore constant witness that the planet really did spin through space. If the Earth stood still, then what could make its waters rush to and fro, rising and falling at regular intervals along the coasts? This view of the tides as the natural consequence of the turning Earth had originally occurred to him nearly twenty years previously, at Venice, when he boarded the barges that carried drinking water into the city from Lizzafusina. Watching the way the large cargoes of water sloshed in response to any changes in the ships' speed or direction, he had found a model for the ebb and flow of the Adriatic and the Mediterranean.

Now, lodged at the Tuscan embassy in the Villa Medici, Galileo passed the early part of January 1616 setting down in writing for the first time his theory of the tides. His social life during this labour consisted of meeting with fifteen to twenty men at a time in the homes of various Roman hosts, where he argued Copernicus's cause in his most compelling style. The nervous Tuscan ambassador, Piero Guicciardini, fairly choked through these evenings, for he dreaded the possible cost of Galileo's actions.

'He is passionately involved in this fight of his', Guicciardini complained to the grand duke, 'and he does not see or sense what it involves, with the result that he will be tripped up and will get himself into trouble, together with anyone who supports his views. For he is vehement and stubborn and very worked up in this matter and it is impossible, when he is around, to escape from his hands. And this business is not a joke, but may become of great consequence, and the man is here under our protection and responsibility.'

Galileo needed the evidence of the tides to support Copernicus because his astronomical findings to date had failed to prove the Earth's motion. It was all very well to argue, as Galileo did, that a rotating, revolving Earth made for a more rational universe – that asking the innumerable, enormous stars to fly daily around the Earth at fantastic speeds was like climbing to a cupola to view the countryside and then expecting the *landscape* to revolve around one's head. Such reasoning, however, said nothing about the way God had actually constructed the firmament.

Even Galileo's discovery of the phases of Venus, which he had dealt as a death blow to the Ptolemaic system, did not constitute proof of the Copernican. The planetary system of Danish astronomer Tycho Brahe could take Venus by the horns and still enable the Earth to remain immobile. According to the Tychonic order, the five planets orbited the Sun, while the Sun – surrounded by Mercury, Venus, Mars, Jupiter and Saturn – circled the stationary Earth. Although Tycho had based this theory on decades of careful observations, Galileo dismissed his plan as even sillier than the Ptolemaic. Since he could not prove the Copernican system by telescope alone, however, he turned to the tides to cement the case. He required the seas to rise to the rescue, not merely of Copernicus's reputation or his own, but to preserve Italy's future scientific pre-eminence and – most important – to protect the honour of the Catholic faith. For if the Holy Fathers banned Copernicus, as rumour predicted they might do at any moment, then the Church would endure ridicule when a new generation of telescopes, probably manned by infidels, eventually uncovered the conclusive evidence for the Sun-centred system.

The waters of the world occupy a moving vessel, Galileo wrote in his 'Treatise on the Tides'. This vast container of water turns on its axis once every day and travels around the Sun once a year. The combination of the two Copernican motions accounts for all tides. The timing and magnitude of specific tides in different locations, however, depend also on many contingent factors, including the extent of each body of water (this was why ponds and small lakes lacked tides), its depth (and consequently the volume of fluid involved), the way it orients itself on the globe (since an east–west waterway like the Mediterranean experienced more dramatic tides than the nearly north–south Red Sea), and its nearness to other bodies of water (which proximity could cause powerful currents and floods, as at the Straits of Magellan where the Atlantic met the Pacific Ocean). Galileo, who never once left Italy, had gathered reports from far and wide to flesh out his explication.

'To hold fast the basin of the Mediterranean and to make the water contained within it behave as it does surpasses my imagination,' Galileo declared, 'and perhaps that of anyone else who enters more than superficially into these reflections.'

But here, again, the fact that Galileo could not account for the tides without moving the Earth did not prove that the Earth moved. What's more, his theory of the tides, though carefully crafted and eminently reasonable, was wrong. Throughout his life he ignored the true cause of the tides, which rise and fall by the pull of the Moon, because he failed to see how a body so far away could exert so much power. To him, the concept of 'lunar influence' smacked of occultism and astrology. Galileo occupied a universe without gravity.* As for the force that made moons orbit planets and planets orbit the Sun in Galileo's cosmology, they might as well have been pushed around by angels.

Kepler, Galileo's German contemporary, made the Moon the centrepiece of his own tidal theory. Kepler's thinking, however, riddled with mystical allusions to the Moon's affinity for water, alienated Galileo's strictly logical mind. (Kepler had even posited intelligent beings on the Moon, as builders of the features observed from Earth.) What's more, Galileo may have had some trepidation about relying on the testimony of a German Protestant.

Galileo presented his manuscript treatise on the tides to one of the newest cardinals in Rome, twenty-two-year-old Alessandro Orsini, a cousin of Grand Duke Cosimo. Galileo wanted Cardinal

---

* His successor, Sir Isaac Newton, born the year Galileo died (1642), dignified the idea of action at a distance in 1687 when he published his law of universal gravitation. In fact, the Moon's gravity would create tides in the Earth's oceans even if the Earth itself did not rotate or revolve.

Orsini to pass the paper on to the current pope, Paul V, whose endorsement might help settle the issue. The young cardinal dutifully delivered the paper, but the sixty-three-year-old pontiff refused to read it. Instead, His Holiness pushed the moment to its crisis by convening expert consultors to decide once and for all whether the Copernican doctrine could be condemned as heretical.

The pope summoned his theological adviser, Roberto Cardinal Bellarmino, the pre-eminent Jesuit intellectual who had served as inquisitor in the trial of Giordano Bruno. Cardinal Bellarmino, the 'hammer of the heretics', had once confided to Prince Cesi of the Lyncean Academy that he personally considered the opinion of Copernicus heretical, and the motion of the Earth contrary to the Bible. (This admission prompted Cesi to wonder whether *De revolutionibus* would ever have been published had Copernicus lived after the Council of Trent, instead of before it.)

Bellarmino knew Galileo from meetings at social occasions over a period of some fifteen years, had viewed Jupiter's moons through his telescope in 1611, and highly respected his achievements, which he could appreciate more than most, having studied astronomy himself at Florence. The only fault Cardinal Bellarmino found with Galileo was the man's insistence on treating the Copernican model as a real-life scenario instead of a hypothesis. After all, there was no proof. The cardinal further opined that Galileo should stick to astronomy in public and not try to tell anyone how to interpret the Bible.

Roberto Cardinal Bellarmino

The Council of Trent, Cardinal Bellarmino took pains to point out, prohibited the interpretation of Scripture contrary to the common agreement of the Holy Fathers – all of whom, along with many modern commentators, understood the Bible to state clearly that the Sun travelled around the Earth. 'The words "the Sun also riseth and the Sun goeth down, and hasteth to his place where he arose, etc." were those of Solomon,' Cardinal Bellarmino wrote,

> who not only spoke by divine inspiration but was a man wise above all others and most learned in human sciences and in the knowledge of all created things, and his wisdom was from God. Thus it is not likely that he would affirm something which was contrary to a truth either already demonstrated, or likely to be demonstrated. And if you tell me that Solomon spoke only according to the appearances, and that it seems to us that the Sun goes around when actually it is the Earth which moves, as it seems to one on a ship that the shore moves away from the ship, I shall answer that though it may appear to a voyager as if the shore were receding from the vessel on which he stands rather than the vessel from the shore, yet he knows this to be an illusion and is able to correct it because he sees clearly that it is the ship and not the shore that is in movement. But as to the Sun and the Earth a wise man has no need to correct his judgment, for his experience tells him plainly that the Earth is standing still and that his eyes are not deceived when they report that the Sun, Moon and stars are in motion.

Galileo was still in Rome in February 1616 when the inevitable happened. At the request of Pope Paul V, who devoted his papacy to promulgating Council of Trent reforms, the cardinals of the Holy Office framed the Copernican argument as two propositions to be voted on by a panel of eleven theologians:

I. The Sun is the centre of the world, and consequently is immobile of local motion.

II. The Earth is not the centre of the world, nor is it

immobile, but it moves as a whole and also with a diurnal
motion.

The unanimous verdict of the panel pronounced the first idea
not only 'formally heretical', in that it directly contradicted Holy
Scripture, but also 'foolish and absurd' in philosophy. The theologians found the second concept equally shoddy philosophically,
and 'erroneous in faith', meaning that although it did not gainsay
the Bible in so many words, it nevertheless undermined a matter
of faith.

The consultors cast their ballots on 23 February and reported
their conclusions to the Holy Office of the Inquisition the following day. Although no public announcement came out of official
chambers, Galileo got a special summons and personal notification
of the outcome almost immediately.

On 26 February, two officers of the Inquisition came to collect
him from the Tuscan embassy. They escorted him to the palace
of Lord Cardinal Bellarmino, who personally met him at the door,
holding his cap, as was his polite custom, and bade Galileo follow
him to his chair. There he told Galileo about the independent
panel's ruling against the Sun's placement at the centre of the
universe. Speaking as the pope's representative, Bellarmino admonished Galileo to abandon defending this opinion as fact. No
record survives of Galileo's spontaneous reaction to this dashing
of all his hopeful efforts, but he doubtless bowed to the cardinal's
command.

Several other people showed up unexpectedly at the cardinal's
house to see Galileo, led by Father Michelangelo Seghizzi, the
Dominican commissary general of the Holy Office of the Inquisition, who had been one of the eleven voting theologians on the
recent panel. He also claimed to speak for the pope, telling Galileo
to relinquish the opinion of Copernicus or else the Holy Office
would proceed against him. Again Galileo acquiesced.

The following week, on 5 March, the Congregation of the Index
published a proclamation that expounded the official position on
Copernican astronomy – namely, that it was 'false and contrary to

Holy Scripture'. The decree also named names and called for action. It suspended Copernicus's book until corrections were made in it, 'so that this opinion may not spread any further to the prejudice of Catholic truth'. It also cited another book, by the Carmelite father Paolo Antonio Foscarini, who had enthusiastically supported Copernicus by quoting chapter and verse from both *De revolutionibus* and the Bible, to show how the two texts could be reconciled. Foscarini fared far worse than

Pope Paul V

Copernicus in the decree, because his book was condemned outright – prohibited and destroyed. Nor did the dismal aftermath end there. The printer in Naples who had published Foscarini's book was arrested soon after the March edict, and Father Foscarini died suddenly in early June, at the age of thirty-six.

Given the specificity of the edict, Galileo saw clearly that only the book attempting to square Copernicus with the Bible had been singled out for the harshest treatment. The two other books cited – that of Copernicus himself and another called *On Job* by Diego de Zuñiga – were merely suspended pending certain deletions and corrections. Galileo's own book, the *Sunspot Letters*, which was also circulating at the time, escaped any mention in the edict, though it strongly supported Copernican astronomy. While Galileo had delved deeply into the Bible and its interpretation with his *Letter to the Grand Duchess Cristina*, this work had not yet been published; his 'Treatise on the Tides' likewise existed in manuscript only.

Having been omitted from the text of the edict, and having escaped any personal censure, Galileo brightened. True, the theory he defended had been condemned, but he emerged free to consider it hypothetically, and to nurture the hope that the decree might one day be repealed. He remained the pre-eminent figure in Italian science, as well as the representative of the Florentine House of Medici. Galileo stayed on in Rome another three months, during which time he met again with Cardinal Bellarmino and spent nearly an hour in a private audience with Pope Paul on 11 March.

'I told His Holiness the reason for my coming to Rome,' Galileo wrote home to the Tuscan secretary of state,

> and made known to him the malice of my persecutors and some of their calumnies against me. He answered that he was well aware of my uprightness and sincerity of mind, and when I gave evidence of being still somewhat anxious about the future, owing to my fear of being pursued with implacable hate by my enemies, he consoled me and said that I might put away all care, because I was held in so much esteem both by himself and by the whole congregation of cardinals that they would not lightly lend their ears to calumnious reports. During his lifetime, he continued, I might feel quite secure, and before I took my departure he assured me several times that he bore me the greatest good will and was ready to show his affection and favour towards me on all occasions.

In the wake of the edict against Copernicus, gossip of heresy and blasphemy continued to smear Galileo's name, though he had not been tried or convicted of any crime. In Venice, word spread that Galileo had been summoned to Rome to account for his beliefs and had now been called to account in the strictest sense. Gossip rumbled through Pisa of how Cardinal Bellarmino had forced Galileo to renounce his beliefs and repent. At the end of May, just before Galileo returned to Florence, he appealed to the cardinal for redress and received this vindicating letter of endorsement:

> We, Roberto Cardinal Bellarmino, having heard that it is calumniously reported that Signor Galileo Galilei has in our hand abjured and has also been punished with salutary penance, and being requested to state the truth as to this, declare that the said Signor Galilei has not abjured, either in our hand, or the hand of any other person here in Rome, or anywhere else, so far as we know, any opinion or doctrine held by him; neither has any salutary penance been imposed on him; but that only the declaration made by the Holy Fathers and published by the Sacred Congregation of the Index has been notified to him, wherein it is set forth that the doctrine attributed to Copernicus, that the Earth moves around the Sun and that the Sun is stationary in the centre of the world and does not move from east to west, is contrary to the Holy Scriptures and therefore cannot be defended or held. In witness whereof we have written and subscribed these presents with our own hand this 26th day of May 1616.

Silenced but exonerated, Galileo confined himself for the next several years to the safe application of his great discoveries, such as using the moons of Jupiter to solve the problem of finding longitude at sea – especially as success might win him the lucrative prize offered by the king of Spain – and studying the companion bodies of Saturn to try to determine their true size and shape.

On 4 October, the feast day of Saint Francis of Assisi, Galileo heard his elder daughter profess her vows at the Convent of San Matteo in Arcetri, about a mile from Florence, where she had already lived for three years. It is possible that when Galileo first arranged for his girls' entry into the convent, he had only their immediate future in mind, and not a lifetime plan. Nevertheless, no husbands had been found.

The form of life of the Order of the Poor Sisters which the blessed Francis founded is this: to observe the holy Gospel of

our Lord Jesus Christ, by living in obedience, without any-
thing of one's own, and in chastity.

THE RULE OF SAINT CLARE, chapter I

At the ceremony of her investiture Virginia relinquished her given name to be known henceforward as Suor Maria Celeste – the name God had chosen for her and whispered in her heart.

From then on, it shall not be permitted her to go outside the
monastery.

THE RULE OF SAINT CLARE, chapter II

The next autumn, on 28 October 1617, Livia followed her sister to become Suor Arcangela. Both girls would spend the rest of their lives at San Matteo.

He Himself deigned and willed to be placed in a sepulchre of
stone. And it pleased Him to be so entombed for forty hours.
So, my dear Sisters, you follow Him. For after obedience,
poverty and pure chastity, you have holy enclosure to hold
on to, enclosure in which you can live for forty years either
more or less, and in which you will die. You are, therefore,
already now in your sepulchre of stone, that is, your vowed
enclosure.

THE TESTAMENT OF SAINT COLETTE

In a desultory manner, Galileo continued to share his abortive theory on the tides with friends in Italy and abroad. 'I send you a treatise on the causes of the tides', Galileo replied in 1618 to a request from Austrian archduke Leopold for a sample of his work, 'which I wrote at the time when the theologians were thinking of prohibiting Copernicus's book and the doctrine enounced therein, which I then held to be true, until it pleased those gentlemen to prohibit the work and to declare the opinion to be false and contrary to Scripture. Now, knowing as I do that it behooves us to obey the decisions of the authorities and to believe them, since they are guided by a higher insight than any to which my humble mind can of itself attain, I consider this treatise which I send you to be merely a poetical conceit, or a dream . . . this fancy of mine . . . this chimera.'

[VIII]

Conjecture here among shadows

GALILEO'S COLLECTED CORRESPONDENCE brims with allusions to illnesses that often kept him from replying sooner to someone or forced him to close a letter in haste. Changes in the weather 'molested' him, his first biographer noted, and he typically fell sick in spring or autumn, or both, about every other year throughout his adult life. Although Galileo rarely elaborated on the nature of these health crises, he may have suffered from some form of relapsing fever contracted during the cave incident in Padua. Or he may have been a victim of malaria or typhoid, a common enough plight in Italy during that period. Another possible explanation for his pattern of repetitive malaise is an unspecified rheumatic disease, possibly gout, which could have accounted for the 'very severe pains and twinges' his biographer said he sustained 'in various parts of his body'. Gout also causes painful kidney stones (when the excess uric acid in the blood, typical of this disease, gets deposited as crystals in the kidneys as well as in the joint spaces), and Galileo complained more than once of prolonged kidney trouble. The quantities of red wine he produced and drank would only have exacerbated the condition (by raising his uric acid level). Even at a time when wine was generally considered the safer alternative to water, doctors recognised the causal connection between alcohol and attacks of gout. Galileo's daughter, who made many of his pills and tonics in the convent apothecary shop, frequently counselled him in her letters to limit 'the drinking that is so hurtful to you' because of the 'great risk of getting sick'.

Other symptoms Galileo sometimes singled out for specific mention included chest pain, a hernia for which he wore a heavy iron truss, insomnia, and various problems with his eyes – particularly unfortunate for an astronomical observer. 'As a result of a certain affliction I began to see a luminous halo more than two feet in diameter around the flame of a candle', Galileo wrote of one such condition to a colleague, 'capable of concealing from me all objects which lay behind it. As my malady diminished, so did the size and density of this halo, though more of it has remained with me than is seen by perfect eyes.' His frequent telescope demonstrations may have predisposed him also to ocular infections, easily communicated by sharing an eyepiece.

After Galileo moved to Florence in 1610, poor health and long periods of recuperation frequently drove him out of the city into the surrounding hills. 'I shall have to become an inhabitant of the mountains,' Galileo vowed while he and his mother and the two little girls still resided on a city street, 'otherwise I shall soon dwell among the graves.'

For several ensuing years he relied gratefully on the hospitality of his friend and follower Filippo Salviati, who rescued Galileo from the foul city air. At Salviati's Villa delle Selve in the hills of Signa, fifteen miles west of Florence, Galileo spent enough time to write the better part of two books – *Bodies in Water* and *Sunspot Letters* – while convalescing from his typical ills. When his ready access to this retreat ended in 1614 with Salviati's death, Galileo pressed the search for his own year-round haven.

In April of 1617, he took a fine villa atop the hill called Bellosguardo ('beautiful sight') on the south side of the River Arno, offsetting the high annual rent of one hundred *scudi* by selling the grain and broad beans grown on the property. From his new eyrie, Galileo enjoyed an unobstructed panorama of the heavens, with a downwards vista that swept the russet roofs, domed churches and city walls of Florence. To the east he could see the olive-green hillside of Arcetri, where his daughters lived inside the walled Convent of San Matteo. It took him three-quarters of an hour on foot or by mule – when he was up to the trip – to visit them.

Despite the salubrious atmosphere at Bellosguardo, however, another serious illness struck Galileo towards the end of 1617 and held him in its grip until spring came. In May 1618, thankful to be freed from his sickbed at last, he set out on a pilgrimage across the Apennine Mountains to the Adriatic coast, where he visited the 'Casa Santa' – the House of the Virgin Mary in Loreto. This former residence of the Blessed Virgin, according to local legend, had abruptly uprooted itself from the Holy Land in the year 1294 and flown on the wings of angels to the laurel grove (*loreto* in Italian) that gave the present town its name. Galileo had first talked of worshipping at the popular shrine in 1616, after he escaped unscathed from the Copernican uproar in Rome, but events and maladies had kept him from fulfilling that intention until now, when he could also offer thanks for his recent recovery and pray for improved health in the future.

He returned home in June to Bellosguardo and to his son, Vincenzio, whom he had brought from Padua in 1612 at the age of seven. By 1618, their male-dominated household also included two new students, Mario Guiducci and Niccolò Arrighetti, who, like Castelli before them and others to come after, would remain Galileo's devoted friends for life. The thirtyish scholars busied themselves all that summer copying the master's early theorems

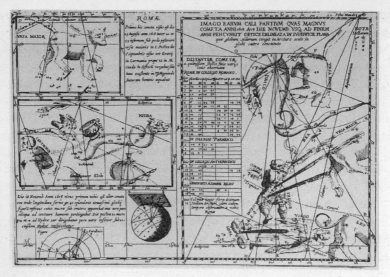

The three comets of 1618

on motion, to help him return to the fundamental work he had
forsaken in 1609 for the telescope. They mined the dense jumble
of his Paduan notes and prepared neat sheets of paper, written
extravagantly on one side only, for his review and revision.

In September, just when Galileo's student assistants had finished
this preliminary work, another bout of illness prevented him from
building on it as planned. The delay might have been merely tem-
porary, except that while Galileo languished, the heavens sent him
a new mystery to ponder, and this apparition initiated a cascade of
events that postponed the publication of his motion studies for
another two decades.

A small comet glowed in the skies over Florence that September
of 1618. Though unspectacular, as comets go, it was nevertheless
the first comet to appear since the birth of the telescope. Other
astronomers took to their rooftops with instruments of Galileo's
design, but Galileo himself remained indoors an invalid. Then
another comet arrived in mid-November, while Galileo unfortu-

nately fared no better than before. And even by the end of November, when a truly brilliant third comet burst on the scene to garner the attention of observers all over Europe, Galileo still could not stand among them.

'During the entire time the comet was visible', he reported later, 'I was confined by illness to my bed. There I was often visited by friends. Discussions of the comets frequently occurred, during which I had occasion to voice some thoughts of mine which cast doubt upon the doctrines that have been previously held on this matter.' In fact, Galileo saw only one important comet his whole life – the big bright one of 1577, in his youth – and never did figure out what these objects really were.

Most of Galileo's contemporaries feared comets as evil omens. (Indeed the three 1618 examples were presently seen, with hindsight, as heralds of the Thirty Years' War, which broke out in Bohemia the same year.) Aristotelian philosophers figured comets for atmospheric disturbances. The fact that comets came and went, changing their fuzzy-glow appearance all the while, automatically relegated them to the sublunar sphere between the Earth and the Moon, where they were thought to be ignited by friction of the spheres' turning against the upper reaches of the air.

It may seem incredible that Galileo resisted the temptation to go outdoors in the autumn of 1618 long enough to view any one of the three comets, especially since he felt well enough to enter into intellectual discussions with visitors. But in fact the November night air held terrible danger for him, a man well past fifty now, who had spent most of the current year battling one malady after another. Moreover, as Galileo no doubt knew from his friends' accounts, he would not have seen much even if he had risked his own study of these objects. A comet, or 'hairy star', retained its blurred contours despite the aid of the most powerful telescope.* Unlike the fixed stars that resolved into points of light when the

* No clear, close-up view of any comet could be obtained until 1986, when several spacecraft observed Halley's comet during its recent return. Images revealed the body to be a dark clump of icy debris – a 'dirty snowball' – that sprouts a great head and tail of glowing gas and dust whenever its highly elliptical orbit carries it near the Sun.

telescope stripped them of their rays, or planets that turned to tiny globes, a comet could not be brought into sharp focus. And Galileo held back because he believed – in agreement for once with his Aristotelian contemporaries, though not for the same reasons – that comets belonged to the Earth's atmosphere.

Galileo thus rejected the findings of his Danish predecessor, Tycho Brahe, who had observed the great comet of 1577 and another in 1585. Tycho, probably the most able naked-eye stargazer who ever lived, followed that comet every night with his oversized measuring instruments to determine its position. It lay beyond the Moon, he discovered through position studies, perhaps as far as Venus, and that meant one of two things to his sixteenth-century way of thinking: either the comet had come crashing through Aristotle's crystalline celestial spheres, or the celestial spheres did not exist. Tycho chose the latter scenario, emboldened by having been the first European, in 1572, to identify a nova, which convinced him that changes *could* occur in the 'immutable' heavens.

Galileo, when he witnessed the next nova in 1604, backed the deceased Tycho's interpretation of the new star's nature and significance. But he despised Tycho's planetary system for its poor compromise between Ptolemy and Copernicus. And as for the comet Tycho had tracked so carefully, Galileo dismissed it as a will-o'-the-wisp. He took comets to be anomalous illuminations in the air – most likely reflections of sunlight bounced off high-altitude vapours – not heavenly bodies *per se*. You could no more gauge the distance to a comet, Galileo believed, than you could catch a rainbow or contain the aurora borealis.

None of the news, notes or queries on the 1618 comets that reached Galileo shook him from his sceptical stance. Nor was he impressed by the pamphlet sent him from Rome containing a comet lecture delivered at the Collegio Romano and published in early 1619. Its author, Jesuit astronomer Father Orazio Grassi, argued on the basis of his studies that the path of the late-November comet carried it between the Sun and the Moon. This was a remarkable conclusion for any Jesuit to reach, because the

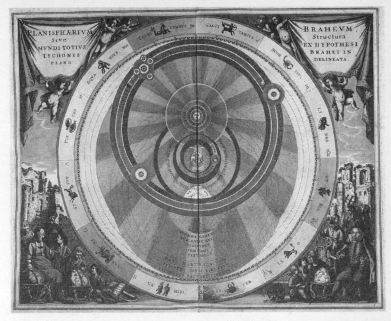

Tycho Brahe's system of the world

Collegio Romano did not dispute Aristotle lightly. Nevertheless, Galileo doubted Father Grassi's distance estimates, just as he had questioned Tycho's, on the grounds that comets had no substance. Father Grassi furthermore committed several mathematical mistakes in his calculations that led him to estimate the volume of the comet, body and 'beard' together, at billions of times the size of the Moon – a ridiculous exaggeration in Galileo's view. Worse, in describing his telescope observations of the comet, Father Grassi exposed his ignorance of the instrument's fundamental principles, inviting Galileo's scorn.

Just at this juncture, Galileo's student Mario Guiducci was elected consul of the prestigious Florentine Academy, which honour obliged him to present a pair of lectures in the spring of 1619. He chose comets as his topic. Galileo wrote much of the content for him, expressing his own bewilderment while negating the work of Tycho and Father Grassi: 'Hence we must be content with what little we may conjecture here among shadows, until there shall be given to us the true constitution of the parts of the universe, inasmuch as that which Tycho promised us still remains imperfect.'

Father Grassi took umbrage at the published version of these talks, which appeared in June 1619 under the title *Discourse on the Comets*. Galileo – for everyone rightly assumed him to be its author – seemed to have singled out the Jesuits as targets of attack: first Father Scheiner (the 'Apelles' of the *Sunspot Letters*) and now Father Grassi – even though the Jesuit Collegio Romano had always upheld Galileo's discoveries and treated him with great respect.

Father Grassi's angry, offended published rebuttal followed swiftly in the book *Libra Astronomica*, or *Astronomical and Philosophical Balance*, which he wrote in Latin under the pen name Lothario Sarsi, a purported student of his. As its title promised, the *Libra* of 1619 hung Galileo's ideas about comets on a steelyard balance scale and found them weightless.

Compelled to respond and silence the noisy barking of his opponents, Galileo began retorting right on the title page of his riposte. He called it *Il Saggiatore*, or *The Assayer* – thus replacing the crude scale of the *Libra* with the more delicate balance assayers used to analyse the quantity of pure gold in gold ore. Father Grassi, retaliating again later in his turn, accidentally on purpose referred to this book as *Assaggiatore*, or *Winetaster* – to imply that Galileo, a notorious lover of good wine, had been drinking when he wrote *The Assayer*.

In 1620, as the tenor of the comet debate turned nastier, the Holy Congregation of the Index raised the spectre of the Edict of 1616 by announcing at last the necessary corrections that must be made to Copernicus's text, *De revolutionibus*, in order to have it removed from the Index of Prohibited Books. The congregation insisted on watering down some dozen statements by Copernicus affirming the Earth's motion, in order to make them sound more like hypothetical suggestions. Galileo dutifully penned the required changes into his own copy of *De revolutionibus*, though he took care to cross out the offending passages with very light strokes.

Galileo ventured no mention of the Copernican theory in *The Assayer*. Such discussion would have been imprudent, given the edict, but also irrelevant: Copernicus had not discussed comets in his book, and Galileo's view of comets as optical illusions automatically divorced them from the order of the Sun and planets as far as he was concerned. He even derided 'Sarsi' and 'his teacher' for granting comets the status of quasi-planets. 'If their opinions and their voices have the power to call into existence the things they have considered and named,' quipped Galileo, 'why then I beg them to do me the favour of considering and naming "gold" a lot of old hardware that I have about my house.'

Indeed, Galileo persisted, the play of the Sun's light could set the most mundane objects aglitter, to fool the unsuspecting: 'Sarsi has but to spit upon the ground and undoubtedly he will see the appearance of a natural star when he looks at his spittle from the point towards which the Sun's rays are reflected.'

Galileo took the occasion of *The Assayer* to mock the philosophical terms that masqueraded as scientific explanations in his day. He noted that *sympathy*, *antipathy*, *occult properties*, *influences* and their like were all too often 'employed by some philosophers as a cloak for the correct reply, which would be: "I do not know."'

'That reply', he reiterated, 'is as much more tolerable than the others as candid honesty is more beautiful than deceitful duplicity.'

Avoiding the forbidden topic of the world system, *The Assayer* thus considered the current comet controversy in the larger context of the philosophy of science. Galileo drew an unforgettable distinction between the experimental method, which he favoured, and the prevailing dependence on received wisdom or majority opinion. 'I cannot but be astonished that Sarsi should persist in trying to prove by means of witnesses something that I may see for myself at any time by means of experiment,' Galileo wrote.

Witnesses are examined in doubtful matters which are past and transient, not in those which are actual and present. A judge must seek by means of witnesses to determine whether Pietro injured Giovanni last night, but not whether Giovanni was injured, since the judge can see that for himself. But even in conclusions which can be known only by reasoning, I say that the testimony of many has little more value than that of few, since the number of people who reason well in complicated matters is much smaller than that of those who reason badly. If reasoning were like hauling I should agree that several reasoners would be worth more than one, just as several horses can haul more sacks of grain than one can. But reasoning is like racing and not like hauling, and a single Barbary steed can outrun a hundred dray horses.

It took Galileo two years to complete *The Assayer*, beset as he was throughout by many family and official matters. Marina Gamba died in February 1619, leaving Galileo's children officially motherless. Having helped both his daughters take the veil, Galileo now atoned for the messy circumstances of his son's birth by getting Grand Duke Cosimo II to legitimise Vincenzio on 25 June, two months before the boy's thirteenth birthday. Cosimo handled this matter-of-factly enough, knowing his own Medici forebears to have fathered at least eight illustrious illegitimate sons, two of whom had become cardinals – and one of those had traded the cardinal's biretta for the pope's tiara as His Holiness Clement VII.

Meanwhile Galileo's mother, Madonna Giulia, grew older and ever grouchier at the house in Florence where she had stayed when her son moved to Bellosguardo. 'I hear with no great surprise that our mother is being so dreadful,' Galileo's brother, Michelangelo, commiserated in October 1619 from the safe distance of Munich. 'But she is much aged, and soon there will be an end to all this quarrelling.'

Madonna Giulia died in September 1620, at eighty-two. Her death was soon followed by the publicly mourned passing of the grand duke, only thirty years of age, in February 1621. Cosimo II, who had come to power at nineteen, now bequeathed the Grand Duchy of Tuscany to the eldest of his eight children, ten-year-old Ferdinando II. The boy also inherited Cosimo's chief mathematician and court philosopher, for Galileo's appointment carried a lifetime tenure. Until Ferdinando reached majority, however, he deferred perforce in all matters to the judgment of his regents: his mother, Archduchess Maria Maddalena of Austria, and his grandmother, the dowager Grand Duchess Cristina of Lorraine.

The necrology of the year 1621 also listed two major figures behind the anti-Copernican edict: Roberto Cardinal Bellarmino, later to be canonised Saint Robert Bellarmine, and Pope Paul V, the founder of the Vatican Secret Archives, where certain papers pertaining to Galileo's 1616 trip to Rome already resided among centuries' worth of private papal documents. Paul V, who had promised to protect Galileo for the rest of his life, died of a stroke on 28 January. A little over a week later, on 9 February, the Sacred College of Cardinals suddenly and unanimously acclaimed his successor, Alessandro Ludovisi of Bologna, as Pope Gregory XV. But the new pontiff's frail health, of which the cardinals had been well aware at the time of his selection, would end his papacy in less than two years.

Galileo fell ill again in early 1621. He recovered by mid-year and completed most of *The Assayer* by the year's end. He composed the long polemic in the form of a letter to his friend and fellow Lyncean in Rome, Virginio Cesarini, Prince Cesi's young cousin, who had been smitten by science under Galileo's influence and written to him during the comet season to offer details of his own cometary observations.

'I have never understood, Your Excellency,' Galileo addressed Cesarini plaintively in the opening pages of *The Assayer*, 'why it is that every one of the studies I have published in order to please or to serve other people has aroused in some men a certain perverse urge to detract, steal or deprecate that modicum of merit which I thought I had earned, if not for my work, at least for its intention.'

In October 1622, Galileo delivered the long-awaited polished manuscript to Cesarini, who fine-tuned it further in collaboration with Prince Cesi before publication. As the printing neared completion the following summer, however, a puff of white smoke from the Sistine Chapel called the process to a halt.

Pope Gregory was dead. Galileo's long-time acquaintance and admirer, Maffeo Cardinal Barberini, now succeeded him as Pope Urban VIII.

Quickly, Prince Cesi created a new engraved title page incorporating the three bees of the Barberini coat of arms. Though Galileo had addressed *The Assayer* to Virginio Cesarini, the Lynceans seized the political expedient of dedicating the book to the new pontiff. They offered up *The Assayer*, which had grown out of a spiteful argument over a trio of comets, as their literary entrée into the papal court of Urban VIII.

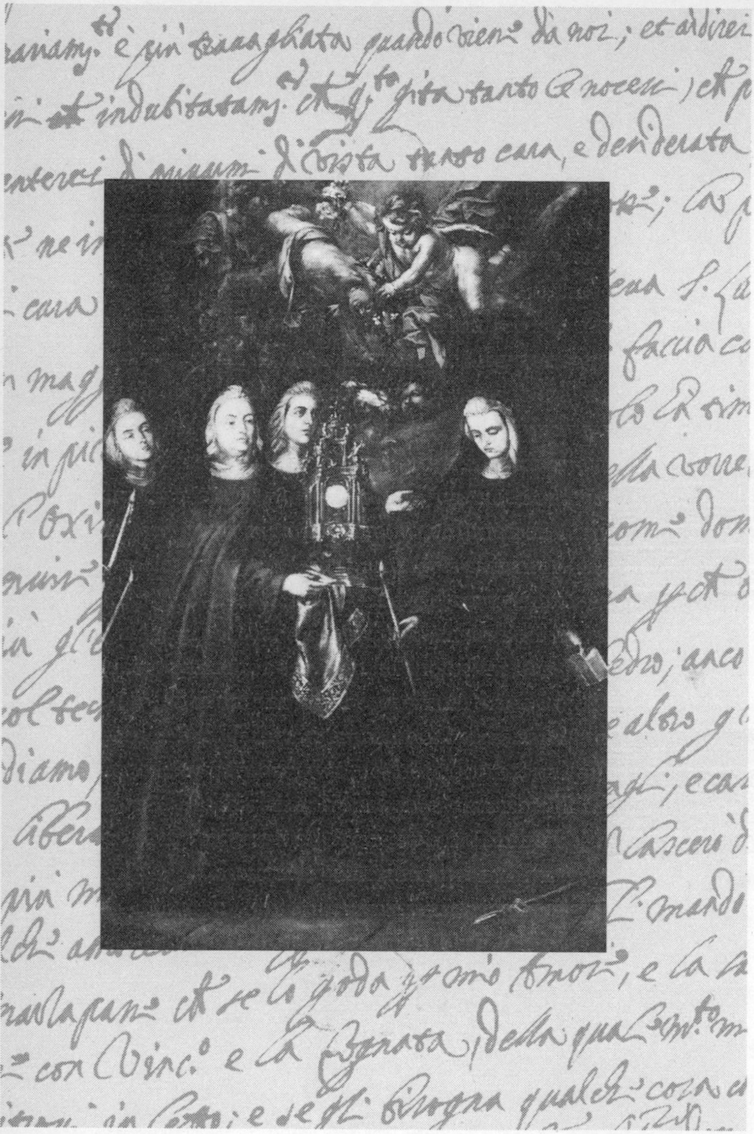

PART TWO

On Bellosguardo

[IX]

How our father is favoured

MOST ILLUSTRIOUS LORD FATHER

THE HAPPINESS I DERIVED from the gift of the letters you sent me, Sire, written to you by that most distinguished Cardinal, now elevated to the exalted position of Supreme Pontiff, was ineffable, for his letters so clearly express the affection this great man has for you, and also show how highly he values your abilities. I have read and reread them, savouring them in private, and I return them to you, as you insist, without having shown them to anyone else except Suor Arcangela, who has joined me in drawing the utmost joy from seeing how much our father is favoured by persons of such calibre. May it please the Lord to grant you the robust health you will need to fulfil your desire to visit His Holiness, so that you can be even more greatly esteemed by him; and, seeing how many promises he makes you in his letters, we can entertain the hope that the Pope will readily grant you some sort of assistance for our brother.

In the meantime, we shall not fail to pray the Lord, from whom all grace descends, to bless you by letting you achieve all that you desire, so long as that be for the best.

I can only imagine, Sire, what a magnificent letter you must have written to His Holiness, to congratulate him on the occasion of his reaching this exalted rank, and, because I am more than a little bit curious, I yearn to see a copy of that letter, if it would please you to show it, and I thank you so much for the ones you have already sent, as well as for the melons which we enjoyed most gratefully. I have dashed off this note in considerable haste, so I beg your pardon if I have for that reason been sloppy or spoken amiss. I send you loving greetings along with the others here who always ask to be remembered to you.

FROM SAN MATTEO, THE 10TH DAY OF AUGUST.
Most affectionate daughter,

S. M. C.

Only four days previously, on 6 August 1623, Maffeo Cardinal Barberini stood sweating in the Sistine Chapel, where he and more than fifty fellow cardinals had been locked together since mid-July in papal election proceedings. In the absence of the unified conviction, inspired by the Holy Spirit, which had led to the *acclamatio* selection of Gregory XV two years before, the Sacred College of Cardinals now prayed for guidance in the balloting. The members

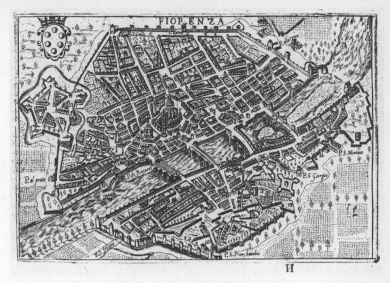

Florence in the late sixteenth century

voted twice a day, morning and afternoon, each one prohibited from endorsing himself, and all of them disguising their handwriting to maintain the secrecy of the selection process. Every time the tabulation failed to produce the required two-thirds majority, the cardinal scrutineers bound the slips of paper and burned them in a special stove with wet straw, which sent up the black smoke of indecision. The typical heat and malaria that visited Rome every summer claimed the lives of six elderly cardinals in the conclave before the assembly at last cast fifty of its fifty-five ballots for Barberini as the new leader of the Church.

'Do you, Most Reverend Lord Cardinal,' the chamberlain asked him face to face, 'accept your election as Supreme Pontiff, which has been canonically carried out?'

'*Accepto*,' he answered with the only word required.

'By what name will you be known?' The choices included his own Christian name, Maffeo, or the name of any pope who had preceded him – except Peter, of course, which tradition forbade.

At a stroke, Maffeo Barberini metamorphosed into Pope

Urban VIII. The ballots, burned with dry straw instead of wet this time, announced their white smoke signal to the crowds around the Vatican, and soon the declaration 'We have a pope!' justified the crescendoing ecstasy of their cries.

Galileo, recognising the potential personal boon of this turn of events, shared his excitement with Suor Maria Celeste by sending her a sheaf of letters that spanned a decade of pleasant exchanges. These dated from Cardinal Barberini's first letter to Galileo in 1611, when he called him a virtuous and pious man of great value whose longevity could only improve the lives of others. The passing years had fanned the cardinal's ardour, so that his 1620 poem, 'Dangerous Adulation', not only cited Galileo as the discoverer of wondrous new celestial phenomena but also used his sunspots as a metaphor for dark fears in the hearts of the mighty. In closing the cover letter to Galileo that accompanied this poetic tribute, the cardinal had signed himself – with noteworthy warmth – 'as your brother'.

Galileo naturally took a more deferential tone in his responses, noting sentiments such as, 'I am your humble servant, reverently kissing your hem and praying God that the greatest felicity shall be yours.' He wrote as often as necessary to keep Cardinal Barberini updated on his scientific work and sent him a copy of each of his books, as well as his unpublished 'Treatise on the Tides', and Guiducci's *Discourse on the Comets*.

The most recent letter from the cardinal bore the date 24 June 1623, just weeks prior to his accession, in which he thanked Galileo for guiding a favourite nephew, Francesco Barberini, through the successful completion of doctoral studies at Pisa.

Galileo wanted nothing so much as to travel to Rome in August to kiss the feet of the new pope and march with his fellow Lyncean Academicians in Urban's investiture procession. The exciting early days of the Barberini pontificate overflowed with promise for all artists and scientists, and particularly those from Urban's native Florence. Striking now, Galileo thought he might well secure the pope's blessing for his own most sensitive projects, and at the same time ensure his son's future.

Galileo had sent Vincenzio off to law school at the University of Pisa the previous year. Now he hoped to obtain a pension for him – a title as a canon that would guarantee Vincenzio a small annual base income for the rest of his life in exchange for little or no real work. Such sinecures influenced the economy of seventeenth-century Italy, often draining the revenues of small dioceses while filling the pockets of absentee landlords, at least half of whom were laymen. Despite the abuses of this system, however, Rome not only

Urban VIII in the first year of his pontificate

tolerated it but also supervised it, because the pensions supported so many more prelates than the papal treasury could afford to pay.

As for Suor Maria Celeste's request to see a copy of her father's congratulatory letter to Urban, Galileo denied it on the grounds that he had not written one. He explained to her that although he enjoyed a most cordial relationship of long standing with His Holiness – enabling him to dedicate his new book to the pope and confidently anticipate a private audience at the Vatican – protocol argued against a personal note on this occasion. Instead, he had proffered his congratulations through the proper channel of Urban's nephew Francesco, who had been Galileo's student.

'It was through your most gentle and loving letter', Suor Maria Celeste confessed a few days later, 'that I became fully aware of my backwardness, in assuming as I did that you, Sire, would perforce write right away to such a person, or, to put it better, to the loftiest lord in all the world. Therefore I thank you for pointing out my error, and at the same time I feel certain that you will, by the love you bear me, excuse my gross ignorance and as many other flaws as find expression in my character. I readily concede that you are the one to correct and advise me in all matters, just as I desire you to do and would so appreciate your doing, for I realise how little knowledge and ability I can justly call my own.'

The date of this letter, 13 August, marked Suor Maria Celeste's twenty-third birthday, although she made no reference to that fact. She reckoned the accumulation of her years now by the anniversary of her nun's vows on 4 October, and by the feast day of the Holy Virgin Mary, whose name she had taken, on 8 September.

Within a week she learned that illness had scotched Galileo's travel plans and forced him into Florence to the home of his late sister, where his recently widowed brother-in-law and his nephew were taking care of him. The news arrived via the steward of San Matteo, a hired man who lived in the convent grounds with his wife, helping the nuns by tending to the heavy work and the traffic with the outside world.

'This morning I learned from our steward that you find yourself ill in Florence, Sire,' she wrote on 17 August, 'and because it sounds to me like something outside your normal behaviour to leave home when you are troubled by your pains, I am filled with apprehension, and fear that you are in much worse condition than usual. Therefore I beseech you to give the steward some account of your state, so that, if you do not fare as badly as we fear, we can calm our anxious spirits. And truly I never resent living cloistered as a nun, except when I hear that you are sick, because then I would like to be free to come to visit and care for you with all the diligence I could muster. But even though I cannot, I thank the Lord God for everything, knowing full well that not a leaf turns without His willing it so.'

Pope Urban VIII suddenly took sick, too. He contracted the fever that ran epidemic through Rome that summer of 1623, and was compelled to postpone his coronation until late September.

Even without the official ceremony, Urban had commenced the exercise of his new powers immediately upon accepting the vote of the conclave, as was his right. On the very day of his election, 6 August, he issued the bulls of canonisation that made saints of Ignatius Loyola and Francis Xavier, the Jesuit founders, and also Philip Neri, 'the Apostle of Rome'.

Barberini
coat of arms

Within weeks, Urban VIII began appointing his brothers and nephews to potent positions in his new regime – inviting detractors to quip that the three bees on the family escutcheon had led former lives as horseflies. The opportunistic Barberini, after all, laid no prior claim to nobility or wealth. Nevertheless, the Barberini pope now commanded the respect of all Christian princes and princes of the Church. Urban turned his married brother, Carlo, into commander-in-chief of the papal armies and made a cardinal of Carlo's erudite eldest son, Francesco. This new cardinal nephew was the same Francesco Barberini who had just earned his doctorate at Pisa, having been a favoured student of Benedetto Castelli and, through him, of Galileo as well. No sooner had the twenty-six-year-old graduate become His Eminence and chief lieutenant to His Holiness than he also found himself elected a member in good standing of the Lyncean Academy.

On 29 September, the day of Urban's coronation, the new pope is said to have displayed the dramatic devotional style that characterised his memorable twenty-year tenure. As he prepared to receive the white papal robes and the velvet Shoes of the Fisherman, he threw himself before the altar in the Chapel of Tears. Prostrate, he prayed for death to take him the moment his pontificate veered from the good of the Church – if such a thing should ever happen, God forbid.

Then Urban consented to be carried in the silk-upholstered *sedia gestatoria*, the capacious portable throne flanked by ostrich feather fans, into Saint Peter's Basilica, where the Sacred College of Cardinals installed him according to the ancient ceremony:

> *Receive this tiara adorned with three crowns; know that thou art the father of princes and kings, victor of the whole world*

under the earth, the vicar of our Lord, Jesus Christ, to whom
be the glory and honour without end.

Compared with the elderly Paul and the sickly Gregory of recent
memory, the fifty-five-year-old Urban cut a youthful, almost military, figure, especially when seen riding on horseback through the
Vatican Gardens. If the pope resembled a general, he showed he
could strategise like one, too. Indeed, history provoked him to use
this talent repeatedly over the next two decades in the waging of
wars for the defence of the Italian peninsula and the integrity
of the Papal States.

Urban would also battle the Protestant Reformation, which still
continued to erode the power of the Roman Church, by stepping
up Catholic Reform measures in his own style. He foresaw
improved ecclesiastical education and networks of foreign missions radiating from a Roman base.

'This is a city upon a hill', Urban pronounced in announcing a
thorough investigation of Rome's own religious health, 'which is
exposed for the whole world to gaze upon.'

Urban intended also to gild and glorify the physical beauty of
the Holy See with new building projects and monuments. Their
construction would employ armies of architects, sculptors and
painters – and invent Urban's reputation as a great patron of the arts.
Upon hearing that a group of admirers had expressed the wish to
commission a monument in the pope's honour during his lifetime,
instead of after his death as was customary, Urban affirmed: 'Let
them. I am not an ordinary Pope either.'

As bookish as he was bold, Urban peopled his curia with literati.
He chose for his Master of Pontifical Ceremonies Monsignor Virginio Cesarini, Lyncean, who wrote acclaimed poetry and had
pursued the study of mathematics after hearing an inspirational
lecture by Galileo. It was from Cesarini that Galileo had received
observations of the 1618 comets, and to Cesarini that Galileo
addressed his ultimate reply to the Jesuit Father Grassi in *The*
Assayer. Immediately upon this book's long-awaited publication in

Rome in late October 1623, Cesarini began reading it aloud to Urban at mealtimes.

Urban, who had studied under the Jesuits at the Collegio Romano, could not help admiring the style with which Galileo skewered Grassi, via 'Sarsi', on his similes: 'I believe that good philosophers fly alone, like eagles,' Galileo said in *The Assayer*, 'and not in flocks like starlings. It is true that because eagles are rare birds they are little seen and less heard, while birds that fly like starlings fill the sky with shrieks and cries, and wherever they settle befoul the earth beneath them.'

Charmed, Urban declared himself all eagerness for Galileo to come to see him. But the great Florentine philosopher remained ill through the autumn, and in the winter the harsh weather restrained him. While Urban awaited Galileo's arrival, he continued to have Galileo's words read aloud at table: 'The crowd of fools who know nothing, Sarsi, is infinite. Those who know very little of philosophy are numerous. Few indeed are they who really know some part of it, and only One knows all.'

Pope Urban found his favourite passage from *The Assayer* in Galileo's parable about the song of the cicada, which demonstrated the boundless creativity of God in the bounty of Nature. 'Once upon a time, in a very lonely place, there lived a man endowed by Nature with extraordinary curiosity and a very penetrating mind,' this story began.

> For a pastime he raised birds, whose songs he much enjoyed; and he observed with great admiration the happy contrivance by which they could transform at will the very air they breathed into a variety of sweet songs.
>
> One night this man chanced to hear a delicate song close to his house, and being unable to connect it with anything but some small bird he set out to capture it. When he arrived at a road he found a shepherd boy who was blowing into a kind of hollow stick while moving his fingers about on the wood, thus drawing from it a variety of notes similar to those of a bird, though by quite a different method. Puzzled, but

impelled by his natural curiosity, he gave the boy a calf in exchange for this flute and returned to solitude. But realising that if he had not chanced to meet the boy he would never have learned of the existence of a new method of forming musical notes and the sweetest songs, he decided to travel to distant places in the hope of meeting with some new adventure.

As the man roved, he encountered songs made by 'a bow . . . sawing upon some fibres stretched over a hollowed piece of wood', by the hinges of a temple gate, by 'a man rubbing his fingertip around the rim of a goblet', and by the beating wings of wasps.

And as his wonder grew, his conviction proportionately diminished that he knew how sounds were produced; nor would all his previous experiences have sufficed to teach him or even allow him to believe that crickets derive their sweet and sonorous shrilling by scraping their wings together, particularly as they cannot fly at all.

Well, after this man had come to believe that no more ways of forming tones could possibly exist . . . he suddenly found himself once more plunged deeper into ignorance and bafflement than ever. For having captured in his hands a cicada, he failed to diminish its strident noise either by closing its mouth or stopping its wings, yet he could not see it move the scales that covered its body, or any other thing. At last he lifted up the armour of its chest and there he saw some thin hard ligaments beneath; thinking the sound might come from their vibration, he decided to break them in order to silence it. But nothing happened until his needle drove too deep, and transfixing the creature he took away its life with its voice, so that he was still unable to determine whether the song had originated in those ligaments. And by this experience his knowledge was reduced to diffidence, so that when asked how sounds were created he used to answer tolerantly that although he knew a few ways, he was sure that many more existed which were not only unknown but unimaginable.

This section of *The Assayer* delighted Urban with its graceful language and poetic conceit, and even more because it expressed his own philosophy of science. To wit: as earnestly as men may seek to understand the workings of the universe, they must remember that God is not hampered by their limited logic – that all observed effects may have been wrought by Him in any one of an infinite number of omnipotent ways, and these must ever evade mortal comprehension.

[X]

To busy myself in your service

GALILEO'S SUMMERTIME ILLNESS OF 1623 is the first of his infirmities to be documented in the surviving letters from his daughter. Although these offer no good clue, unfortunately, to the specific nature of Galileo's disorder, they demonstrate clearly Suor Maria Celeste's familiarity with his indifferent health, and how it preoccupied her. She hung on word of him, which arrived now from the steward, now from her uncle Benedetto Landucci. The first week of Galileo's stay in Florence she prepared him a treat of marzipan shaped like little fish, and the second week, hearing he could hardly eat a bite of anything, she found four fresh plums to tempt his appetite. After he returned to Bellosguardo in September, she willingly helped him with his correspondence.

Galileo's affliction compromised his handwriting, so that his extant papers include many documents penned in a small, cramped hand, with the lines on a steep slant up or down the page, as though their author were working lying down. (Galileo's followers used his changeable penmanship as a clue to help establish chronological order among the welter of his undated papers, relying also on ink colour, for he purchased different types in different cities, and any special characteristics of paper, such as the distinguishing rhinoceros watermark from the end of his Paduan period.)

Although Suor Maria Celeste often sacrificed neatness for speed in her own writing, she readily lent her stylish script to Galileo, as well as to the mother abbess, for whom she did not merely copy letters from drafts but composed them from scratch.

To begin, she enlarged the first letter of the first paragraph's first word, as in an illuminated manuscript, and festooned the capital with loops and tails, turning her pen to vary the width of her strokes from a hair's breadth to a broad ribbon. As her sentences flowed along to the right, she bent all the lower-case *d*s back over themselves in the opposite direction, hiding each small *d* circle beneath a huge canopy that might shade an entire word. Where the tip of a tall right-tilting letter met the top of a left-slanting *d*, she joined them in a pointed arch. Occasionally she decorated the heads of pages and the ends of paragraphs with flourishes, inserted extra ripples of squiggles into salutations and sign-offs, and on her envelopes she drew, more than wrote, the neighbourhood – Bellosguardo – or city name – Rome, Florence, Siena – that sufficed for an address.

'Here is the copied letter, Sire,' Suor Maria Celeste complied with Galileo's request on 30 September, 'along with the wish that it meet with your approval, so that at other times I may again be able to help you by my work, seeing as it gives me such great pleasure and happiness to busy myself in your service.'

She soon found another way to assist him, by sewing a set of table linens her father could use on his trip to Rome – for carrying food or spreading over tables at inns along the way – even though the press of her convent duties left her little if any free time. In

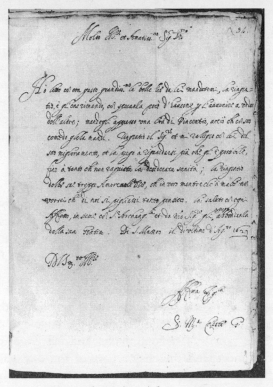

Maria Celeste's letter of 31 August 1623

addition to praying the Divine Office (the Liturgy of the Hours), the major daily occupation of the cloistered nuns, the Poor Clares at San Matteo worked long hours to sustain their ever-struggling economy. They grew a few fruits and vegetables to feed themselves, did all their own cleaning and cooking, and also produced articles for outside sale, such as fine embroidered handkerchiefs, lace, herbal medicines, and bread in the summertime, when it was too hot for anyone else to bake. The rough brown habits they wore, with black linen veil and knotted cord belt, never showed the dirt of their menial labours.

Suor Maria Celeste, musically talented like most members of her family, also directed the choir from time to time and taught the novices how to sing Gregorian chant. In her capacity as the convent's apothecary, she assisted the visiting doctor, fabricated remedies in pill or tonic form, and nursed the sick nuns in the infirmary, where Suor Arcangela often occupied a bed. Although Suor Maria Celeste, who spoke for both of them in her letters, never directly accused Suor Arcangela of malingering, she some-times alluded to a hysterical component in her sister's complaints. The younger daughter's moodiness and taciturnity may have characterised her nature from childhood. Or the traits may have developed in reaction to cloistered life, which had been her father's choice and not her own.

The arduous existence of the Poor Clares was described baldly by a contemporary of Suor Maria Celeste's and Suor Arcangela's – Maria Domitilla Galluzzi, who entered the house of the Clarisses in Pavia in 1616, and later wrote her own interpretation of the Rule of Saint Clare.

'Show her how we dress in vile clothing,' Maria Domitilla coun-selled any nun introducing a candidate for admission to the sisters' way of life, 'always go barefoot, get up in the middle of the night, sleep on hard boards, fast continually, and eat crass, poor and lenten food, and spend the major part of the day reciting the Divine Office and in long mental prayers, and how all of our re-creation, pleasure and happiness is to serve, love and give pleasure to the beloved Lord, attempting to imitate his holy virtues, to mortify and vilify ourselves, to suffer contempt, hunger, thirst, heat, cold and other inconveniences for his love.'

Although the Council of Trent had specifically denounced the once common practice of forcing young women to take the veil, the percentage of patrician girls entering convents in Florence actually rose through the remainder of the sixteenth century and continued to increase on into the seventeenth. Whether or not Galileo's daughters walked willingly into San Matteo, Suor Maria Celeste found her place there. The same cannot be said with cer-tainty of Suor Arcangela. If she ever wrote a letter to her father, it has not been preserved.

On top of all her appointed tasks, Suor Maria Celeste voluntarily took up the sort of domestic work she would have performed for her father and brother had she lived in their home on Bellosguardo. Despite her distance, she insinuated herself between the two of them, always placating her father while pleading Vincenzio's point of view in matters large or small.

'Vincenzio stands in dire want of new collars', she informed Galileo in an undated letter,

> even though he may not think so, as it suits him to have his used ones bleached as the need arises; but we are struggling to accommodate him in this practice, since the collars are truly old, and therefore I would like to make him four new ones with lace trim and matching cuffs; however, since I have neither the time nor the money to do this all by myself, I should like for you to make up what I lack, Sire, by sending me a *braccio* of fine cambric and at least 18 or 20 *lire* to buy the lace, which my lady Ortensa makes for me very beautifully; and because the collars worn nowadays tend to be large, they require a good deal of trimming for properly finishing them; moreover, seeing as Vincenzio has been so obedient to you, Sire, in always wearing his cuffs, I maintain, for that reason, he deserves to have handsome ones; and therefore do not be astonished that I ask for this much money.

The wide white collars Suor Maria Celeste sewed, washed and bleached for her father and brother frame Galileo's face in every formal portrait painted of him. At home alone, however, tending his experiments or his grapevines, the shirtsleeved Galileo donned an old leather apron.

'I am ashamed that you see me in this clown's habit,' he reportedly said to a group of distinguished visitors who arrived one afternoon to find him in his garden, in his work clothes. 'I'll go and dress myself as a philosopher.'

But he must have been jesting, for when the men asked Galileo why he didn't hire someone to take over his manual labour, he replied: 'No, no; I should lose the pleasure. If I thought it as much fun to have things done as it is to do them, I'd be glad to.'

This outdoor recreation tempered the concentration of Galileo's scholarly work and kept him close to Nature. Though the time he spent in his kitchen garden, his orchard and his vineyard restored his spirits, the hours took their toll on Galileo's well-worn work garment, which he periodically sent to Suor Maria Celeste for repair.

MOST ILLUSTRIOUS AND BELOVED LORD FATHER

I AM RETURNING THE rest of your shirts that we have sewn, and the leather apron, too, mended as best I could. I am also sending back your letters, which are so beautifully written that they have only kindled my desire to see more examples of them. Now I am tending to the work on the linens, so that I hope you will be able to send me the trim for borders at the ends, and I remind you, Sire, that the trimmings needs to be wide, because the linens themselves are rather short.

I have just placed Suor Arcangela once more into the doctor's hands, to see, with God's help, if she can be relieved of her wearisome illness, which causes me no end of worry and work.

Salvadore [Galileo's servant] tells me that you want to pay us a visit soon, Sire, which is precisely what we so desire; though I must remind you that you are obliged to keep the promise you made us, that is, to spend an entire evening here, and to be able to have dinner in the convent parlour, because we deliver the excommunication to the tablecloth and not the meals thereon.

I enclose herewith a little composition, which, aside from expressing to you the extent of our need, will also give you the excuse to have a hearty laugh at the expense of my foolish writing; but because I have seen how good-naturedly you always encourage my meagre intelligence, Sire, you have lent me the courage to attempt this essay. Indulge me then, Lord Father, and with your usual loving tenderness please help us. I thank you for the fish, and send you loving greetings along with Suor Arcangela. May our Lord grant you complete happiness.

FROM SAN MATTEO, THE 20TH DAY OF OCTOBER 1623.
Most affectionate daughter,

S. M. C.

Suor Maria Celeste's casual reference to excommunication poked private fun at a practice of the Poor Clares. The Rule of the order stated plainly that no visitors could enter the refectory where the nuns dined. The Convent of San Matteo, however, maintained a separate parlour where a sister's family members might properly be received. They could bring their own food, too, and share it with her. Thus the dishes themselves, whether cooked in the convent or carried in by the guests, could be eaten with impunity, so long as everyone ate in his or her proper place. A black iron grille separated the parlour from the nuns' quarters, and all exchanges passed through the lattice of its bars. Another grille pierced the wall near the altar in the adjacent Church of San Matteo, so that the voices of the nuns singing in their choir could reach the townspeople attending mass on the other side. Although the Poor Clares devoted their earthly lives to praying for all the souls of the world, they required the maintenance of a severed space for this work, where they lived hidden in God's embrace.

These practices traced back to the early thirteenth century, when Francis of Assisi spurned splendid wealth to found his Order of Friars Minor on the principles of poverty, obedience and devotion. The rich, privileged young Chiara Offreduccio, or Clare, joined him as his first female follower in the spring of 1212. Francis cut off her golden hair and sent her begging in the streets of Assisi. In time the two divided their labours so that Francis travelled widely preaching the Gospel, while Clare headed the contemplative second order of Franciscans, known as the Clarisses, or Poor Ladies – the Poor Clares. Clare sequestered herself for life in the convent Francis built her at San Damiano, where she slept on the floor and ate practically nothing. She also initiated the tradition of work in the convents, filling the hours between the daily offices with spinning and embroidery.

> *The Sisters to whom the Lord has given the grace of working should labour faithfully and devotedly after the hour of Terce at work which contributes to integrity and the common good . . . in such a way that, while idleness, the enemy of the soul, is banished, they do not extinguish the spirit of holy prayer and dedication to which all other temporal things should be subservient.*
>
> THE RULE OF SAINT CLARE, chapter VII

By the time Suor Maria Celeste joined the order, the Rule of Saint Clare had relaxed on some issues and tightened on others, as dictated by Church policy and the individual interpretation of each convent's mother abbess. Seventeenth-century parents, for example, paid dowries for their daughters to enter Clarisse convents in Italy – a requirement that might have horrified Francis and Clare. Poverty remained a central tenet of the Rule, rendering all Poor Clares dependent on alms. Suor Maria Celeste perforce appealed frequently to her father for financial help, though she found this duty embarrassing. It was one thing to request money to buy presents for Vincenzio and quite another to ask anything for herself. In the 'little composition' she apologetically enclosed with her letter of 20 October, she apparently meant to soften her current plea by making a comedy of the convent's neediness.

Saint Clare

Unfortunately, the attachment has disappeared, so it is impossible to say whether it took the form of an essay or perhaps a dramatic action of the sort Galileo used to write – and liked to see performed, through the grille, when he visited San Matteo. The nuns also wrote plays, in keeping with convent traditions, for the Church authorities encouraged them to stage spiritual comedies and tragedies drawn from biblical themes as part of their education and recreation.

Whatever Suor Maria Celeste's request, Galileo never failed to fulfil it, precipitating a flood of appreciation in return. Suor Maria Celeste's word for the loving indulgence that characterised her father's attentiveness – *amorevolezza* – appears more than twenty times in her 124 surviving letters, thanking him for some recent act of thoughtfulness or generosity towards herself, her sister, or someone else in the convent. Thus, all the while that Galileo was inventing modern physics, teaching mathematics to princes, discovering new phenomena among the planets, publishing science books for the general public, and defending his bold theories against establishment enemies, he was also buying thread for Suor Luisa, choosing organ music for Mother Achillea, shipping gifts of food, and supplying his homegrown citrus fruits, wine and rosemary leaves for the kitchen and apothecary at San Matteo.

'If I wanted to attempt to thank you with words, Sire, for these recent presents you sent us,' she wrote on 29 October 1623, the week after dispatching her 'little composition', 'I could not imagine how to begin fully to express our indebtedness, and what is more, I believe that such a display of gratitude would not even please you, for, as kind and good as you are, you would prefer true thankfulness of the spirit from us over any demonstration of speeches or ceremonies. We will therefore serve you better if we apply what we do best, and by that I mean prayer, in seeking to recognise and make recompense for this and all the other innumerable, and even far greater gifts that we have received from you.'

Anxiously anticipating her father's imminent departure for Rome, she feared that the separation would be a long one, and she dreaded being deprived of his attentions. As usual, she raised Vincenzio's agenda: 'I want to offer you a good word on behalf of our poor brother, although I may be speaking out of turn, yet I beseech you to forgive him his mistake this time, blaming his youth as the real cause for his committing such a blunder, which, being his first, merits pardon: I therefore entreat you once again to take him with you to Rome, and there, where you will not lack for opportunities, you can give your son the guidance that your paternal duty and all your natural goodness and loving tenderness seek to provide him.'

Vincenzio, now seventeen, had turned out to be the sullen, thankless opposite of his industrious elder sister. Attending college in Pisa, he squandered money and taxed the goodwill of Galileo's dear friend Benedetto Castelli, to whose care he had been entrusted. None of Castelli's letters clarify, however, what particular infraction of Vincenzio's provoked the paternal anger on this occasion.

'For the future', Galileo wrote in exasperation to Castelli of Vincenzio's frequent financial demands during his student days, 'he is to be content with 3 *scudi* a month for pocket money. With this he can buy plaster figures, pens, paper or anything else he likes; and he may consider himself lucky to have as many *scudi* as I at his age had groats.'

Towards the end of November, after *The Assayer* appeared in print and won acclaim in Rome, Suor Maria Celeste requested a copy, among other favours from her father, on what she suspected to be the eve of his departure.

Most Illustrious Lord Father

BETWEEN THE INFINITE LOVE I bear you, Sire, and my fear that this sudden cold, which ordinarily troubles you so much, may aggravate your current aches and indispositions, I find it impossible to remain without news of you: therefore I beg to hear how you are, Sire, and also when you think you will be setting off on your journey. I have hastened my work on the linens, and they are almost finished; but in applying the fringe, of which I am sending you a sample, I see I will not have enough for the last two cloths, as I need almost another four *braccia*. Please do everything you can to get this to me quickly, so I can send them all to you before you leave; as it is for the purpose of your upcoming trip that I have gone to such lengths to finish them.

Since I do not have a room where I can sleep through the night, Suor Diamanta, by her kindness, lets me stay in hers, depriving her own sister of that hospitality in order to take me in; but the room is terribly cold now, and with my head so infected, I cannot see how I will be able to stand it there, Sire, unless you help me by lending me one of your bed hangings, one of the white ones that you will not need to use now while you are away. I am most eager to know if you can do me this service. And another thing I ask of you, please, is to send me your book, the one that has just been published, so that I may read it, as I am longing to see what it says.

Here are some cakes I made a few days ago, hoping to give them to you when you came to bid us adieu. I see that this will not happen quite as soon as I feared, and so I want you to have them before they turn hard. Suor Arcangela continues still to purge herself, and she does not feel terribly well after having had the two cauteries on her thighs. I am still not very well either, but by now I am so accustomed to poor health that I hardly think about it, seeing how it pleases the Lord to keep testing me always with some little pain or other. I thank Him, and I pray that He grant you, Sire, the greatest possible well-being in all respects. And to close I send you loving greetings from me and from Suor Arcangela.

FROM SAN MATTEO, THE 21ST DAY OF NOVEMBER 1623.
Most affectionate daughter,

S. M. Celeste

If you have collars to be bleached, Sire, you may send them to us.

[XI]

What we require above all else

IN THE GREAT SILENCE that descended over the convent after the evening prayers, the thirty sisters of San Matteo lay asleep in their beds, fully clothed. Should Death come to call for one of them in the night, she would be dressed and ready to enter the next life. Or, when the bell summoning the nuns to Matins disturbed the darkness at midnight, they all could rise from their straw mattresses and go at once, without delay, padding in barefoot procession to the choir to meet their bridegroom Jesus by candlelight.

'*Venite adoremus*,' they chanted as they took their places near the altar under the church grille for the Office of the Annunciation

– the first round of devotions in a new day. The nuns touched their foreheads to the stone floor, made the sign of the cross on their lips, then rose and kneeled in worship, supplicating for all those who might be aided by their prayers.

In the pre-dawn shadows following the choral recitation of Matins and Lauds, the sisters returned briefly to their cells for the remainder of their sleep. But it was probably during this period before sunrise on 10 December 1623 that Suor Maria Celeste found the time to complete a secret letter to her father regarding a matter of supreme importance to the convent.

Galileo, still intending to go to Rome, had offered to petition the pope on behalf of the nuns of San Matteo. He wanted Suor Maria Celeste, given her inside knowledge of their plight, to inform him of their greatest need. Over the past few days, she had sought the advice of the mother abbess and other trusted sisters, finally relying on her own observations to reach a decision as to what would serve the community best.

She had rejected the notion of a gift of alms. Of course the convent was indigent, and as a result the nuns often went hungry, but Poor Clares consciously chose to live in continual abstinence. Their founding mother, Clare, herself had cherished the Privilege of Poverty as the clearest imitation of Christ. After her protector, Francis of Assisi, died in 1226, Clare battled successfully to maintain her right to own nothing, despite the objections of Church officials who feared she would starve to death. Bending to her will, Pope Gregory IX permitted the Poor Ladies at San Damiano to continue their tradition of corporate and individual poverty.

> *The Sisters shall not appropriate anything to themselves, neither a house nor a place nor anything whatsoever; and as strangers and pilgrims in this world, serving the Lord in poverty and humility, let them confidently send for alms.*
>
> THE RULE OF SAINT CLARE, chapter VIII

Pope Gregory, however, weighing the general welfare of the Clarisses throughout Italy and France, had forced other convents to accept real property that could be sold or rented for profit. By the seventeenth century, this practice enjoyed wide vogue. Surely the new Pope Urban, if approached by an individual he admired as much as he did Galileo, might easily cede to San Matteo a prosperous estate that generated enough income to buy food and blankets aplenty.

Something far worse than material poverty, in Suor Maria Celeste's analysis, threatened to undermine San Matteo. Therefore she made bold to name the problem to her father, and to suggest precisely how the pope might crush the menace. Her letter described the impropriety of the men attending certain supervisory needs of the convent, beginning with the spiritual guidance that had once come from the blessed Francis himself.

> *Our Visitor shall always be from the Order of Friars Minor according to the will and instruction of our Cardinal. And let him be such as to be well known for his integrity and manner of living. His duty will be to correct faults committed against the form of our profession, whether in the head or in the members.*
>
> THE RULE OF SAINT CLARE, chapter XII

If the visitor was wanting, the chaplains were worse. Yet, since the Church did not allow women to administer the Sacraments, the convent required a chaplain or two for special purposes such as hearing confession, anointing the sick and celebrating the mass.

The current chaplain attending San Matteo exemplified the crowd of uneducated, unethical clerics who infiltrated ecclesiastical ranks at almost every level in seventeenth-century Italy, making a mockery of religious life wherever they came to roost. Although the Council of Trent, at its conclusion in 1563, had tried to rout these unscrupulous characters by calling for the establishment of seminaries – one in every diocese – to train truly devout young men for holy office, such schools had not yet proliferated widely enough, and many incompetent priests continued to obtain

Saint Clare facing the Saracens

positions through apprenticeships or political connections. Suor Maria Celeste surmised that replacing the current priest, who had no experience of monastic life, with a friar would transform morale at the convent.

The Council of Trent had also reduced the tenure of mother abbesses from a lifetime appointment to an elected post that passed to a new occupant by a vote held every three years. At San Matteo later this very day, in fact, just such a ballot would decide the successor to Suor Laura Gaetani, or 'Madonna', as the nuns called her, who was now conducting the Morning Office for the very last time.

Another pealing of the sacristan's bell sounded the beginning of Prime. Fully awake, washed, and reassembled in the choir, the sisters pressed on with their intercessions for the sinful and the anguished. They had not spoken any words of ordinary conversation since the previous evening, following the Rule's admonition to keep silence until the hour of Terce. Now they sang hymns to Suor Maria Grazia's accompaniment at the old, dilapidated organ, breakfasted on bread, and bent to their work.

Some time during the three hours of chores after Terce, before the midday chanting of Saint Clare's own favourite canonical hours of Sext and None, Suor Maria Celeste could have found time to fold her two densely filled sheets of paper lengthwise in quarters, then in half to make a small square, seal the edges, and pass the packet to the steward for delivery to Bellosguardo.

Most Illustrious and Beloved Lord Father

I was hoping to be able to respond in person, Sire, to everything you said in your most solicitous letter of several days ago. I see, however, that time may prohibit us from meeting before you take your leave, and so I am resolved to share my thoughts with you in writing. Above all, I want you to know how happy you made me by offering so lovingly to help our convent. I conferred with Madonna and other elders here, all of whom expressed their gratitude for the nature of your offer; but because they were uncertain, not knowing how to come to a decision among themselves, Madonna wrote to our Governor, and he answered that, since the convent is so impoverished, alms were probably needed more than anything. Meanwhile I had several discussions with one particular nun, who seems to me to surpass all the others here in wisdom and goodwill; and she, moved not by passion or self-interest but by sincere zeal, advised me, indeed beseeched me, to ask you for something which would undoubtedly be of great use to us and yet very easy for you, Sire, to obtain: that is, to implore His Holiness to let us have for our confessor a Regular or Brother in whom we can confide, with the possibility that he may be replaced every three years, as is the custom at convents, by someone equally dependable; a confessor who will not interfere with the normal observances of our Order, but simply let us receive from him the Holy Sacraments: it is this that we require above all else, and so much so that I can hardly express its crucial importance, or the background circumstances that make it so, although I have tried to list several of them in the enclosed paper that I am sending along with this letter.

But because I know, Sire, that you cannot, on the basis of a simple word from me, make such a demand, without hearing from others more experienced in such matters, you can look for a way, when you come here, to broach the question with Madonna, to try to get a sense of her feelings on the matter, and also to discuss it with any of the more elderly mothers, without, of course, exposing your reasons for mentioning such things. And please breathe not a word of this to Master Benedetto [her uncle, the father of Suor Chiara], since he would undoubtedly divulge it to Suor Chiara, who would then spread it among the other nuns, and thus ruin us, because it is impossible for so many brains to be of one mind; and as a consequence the actions of a single person who might be particularly displeased by this idea could thwart our efforts. Surely it would be wrong to let two or three individuals deprive everyone in the group of all the benefits, both spiritual and practical, that could accrue from the success of this plan.

Now it is up to you, Sire, with your sound judgment, to which we appeal, to determine whether you deem it appropriate to pose our entreaty, and how best to present it so as to achieve the desired end most easily; since, as far as I am concerned, our petition seems entirely legitimate, and all the more so for our being in such dire straits.

I made it a point to write to you today, Sire, as this is rather a quiet time, and I think the right time for you to come to us, before things get stirred up again, so that you can see for yourself what may need to be done in respecting the stature of the older nuns, as I have already explained.

Because I fear imposing on you too heavily, I will leave off writing here, saving all the other things to tell you later in person. Today we expect a visit from Monsignor Vicar, who is coming to attend the election of the new Abbess. May it please God to see the one who bends most to His will elected to this post, and may He grant you, Sire, an abundance of His holy grace.

FROM SAN MATTEO, THE 10TH DAY OF DECEMBER 1623. Most affectionate daughter,

S. M. Celeste

The first and foremost motive, which drives us to make this plea, is the clear recognition and awareness of how these priests' paltry knowledge or understanding of the orders and obligations that are part of our religious life, allow us, or, to say it better, tempt us to live ever more loosely, with scant observance of our Rule; and how can one doubt that once we begin to live without fear of God, we will be subject to continual misery with regard to the temporal matters of this world? Therefore we must address the primary cause, which is this one that I have just told you.

A second problem is that, since our convent finds itself in poverty, as you know, Sire, it cannot satisfy the confessors, who leave every three years, by giving them their salary before they go: I happen to know that three of those who were here are owed quite a large sum of money, and they use this debt as occasion to come here often to dine with us, and to fraternise with several of the nuns; and, what is worse, they then carry us in their mouths, spreading rumours and gossiping about us wherever they go, to the point where our convent is considered the concubine of the whole Casentino region, whence come these confessors of ours, more suited to hunting rabbits than guiding souls. And believe me, Sire, if I wanted to tell you all the blunders committed by the one we have with us now, I would never come to the bottom of the list, because they are as numerous as they are incredible.

The third thing will be that a Regular must never be so ignorant that he does not know much more than one of these types, or if he does not know, at least he will not flee the convent, as has been the constant practice of our priests here, on the occasion of any little happenstance, to seek advice from the bishopric or elsewhere, as though that were any way to comport oneself or counsel others; but rather he will consult some learned father of his own Order. And in this fashion our affairs will be known in only one convent, and not all over Florence, as they are now. More than this, if he has gained nothing else from his own experience, he will well understand the boundaries that a Brother must respect between himself and the nuns, in order for them to live as quietly as possible; whereas a priest who comes here without having, so to speak, knowledge of nuns, may complete the whole designated three years of his required stay without ever learning our obligations and Rule.

We are not really requesting fathers of one religious order in preference over another, trusting ourselves to the judgment of he who will obtain and grant us such a favour. It is very true that the Reformed Carmelites of Santa Maria Maggiore, who have come here many times as special confessors, have served us most satisfactorily in the offices we are prohibited from performing ourselves; and I believe that they would better conform to our need. First, being themselves very devout fathers and highly esteemed; and moreover, because they do not covet fancy gifts, nor concern themselves (being well accustomed to poverty) with a grandiose lifestyle, as members of some other Orders have sought here; certain priests sent to us as confessors spent the whole three years serving only their own interests, and the more they could wring out of us, the more skilful they considered themselves.

But, without straining to make further allegations, Sire, I urge you to judge for yourself the conditions at other convents, such as San Jacopo and Santa Monaca, now that they have come under the influence of Brothers who took steps to set them on the proper path.

We are by no means asking to shirk the obedience of our Order, but only to be administered the Sacraments and governed by persons of experience, who appreciate the true significance of their calling.

Legend held that Mother Clare often chanted Sext and None, which commemorate the crucifixion and death of Jesus, in tears. Her daughters at San Matteo followed these offices with prayers of gratitude for the convent's benefactors, and then filed two by two, still singing, to the refectory. There the nun whose turn it was to read aloud at dinner regaled her sisters with stories from the lives of the saints, though not for long, as it took only a few minutes to finish the scant fare, usually broth and a vegetable, before returning to their prayers.

At Vespers in the early afternoon, the nuns knelt in the choir stalls, listening to the bells ring evensong. Another bell, the *capitolo*, rang soon afterwards, during what would normally have been another period of silent work, calling them to chapter for the election that saw Suor Ortensia del Nente, the convent's expert lace maker, inaugurated as their new mother abbess.

> *The one elected should reflect upon what kind of burden she has taken up and to whom an account of the flock entrusted to her is to be rendered. She should also strive to lead the way for the others more by virtue and a holy way of acting than by her office so that roused by her example the Sisters might obey her more out of love than out of fear.*
>
> THE RULE OF SAINT CLARE, chapter IV

The founding Clare, in contrast, had headed San Damiano all her days. These passed mostly in quiet reverence, memorably broken by an invasion of mercenary soldiers in September of 1240, during which skirmish Clare, who had been bedridden by illness for six years, stood up with the assistance of two nuns and drove out the enemy by the power of her prayers. Clare's canonisation in 1255 resulted partly from this act of valour, as well as one supporting nun's eyewitness testimony under oath that God had spoken to Clare – 'I will guard you always and defend you,' He said – while the Saracens scaled the walls of San Damiano, and partly on the basis of miracles following her death as signs of her sanctity. Pilgrims brought to Clare's tomb were variously cured there of epilepsy, paralysis, withered limbs, hunchback, dementia, madness and blindness.

> *I bless you during my life and after my death as much as I am able and even more than I am able, with all the blessings by which the Father of mercies has blessed and will bless his spiritual sons and daughters in Heaven and on Earth. Amen.*
>
> BLESSING OF SAINT CLARE

The days at San Matteo drew to a close during evening meditation, after the hymn and the Rosary, in solitude. Then all the nuns fell to their knees and fully prostrate on the floor to beg each other's forgiveness for any pain one sister might inadvertently have caused another through all the preceding hours.

At Compline they convened chorally once more in the gathering darkness, as each nun prepared to meet her holy bridegroom or her death – whichever the Great Silence held in store for her this night.

[X I I]

Because of our zeal

IN THE SPRING OF 1624, after the cold rain stopped falling and the roads again became passable, Galileo set out at last for the Vatican. On 1 April, he left Florence in a horse-drawn litter lent him for the occasion by the thirteen-year-old grand duke of Tuscany, Ferdinando II. Although Galileo still enjoyed the privileges of his ties to the Pitti Palace and continued to draw his stipend from the University of Pisa as the Medici family's chief mathematician and philosopher, none of these conditions prohibited his courting the greater favour of the new pope. It could even be argued that Galileo required the pope's aegis, in order to guarantee the loyalty of Ferdinando's pious female regents, Madama Cristina and Archduchess Maria Maddalena.

En route to Rome, Galileo stopped for two weeks in Acquasparta as the guest of his friend and patron Prince Cesi, whom he hadn't seen in eight years. Cesi preferred to spend as much time as

Prince Federico Cesi

possible at his country estate, where he happily devoted himself to botany, writing and the compilation of his encyclopedia of natural history – the crowning (though ever unfinished) work of the Lyncean Academy.

Prince Cesi naturally anticipated publishing Galileo's own new book project – the one about the system and composition of the universe – the one the court philosopher had dreamed on and off of writing since his first telescopic observations in Padua some fifteen years before. The Edict of 1616 had all but killed the idea until now, when the vibrancy of Rome's new order promised unprecedented intellectual freedom. First, of course, Galileo had to broach the proposal with Pope Urban and gauge his response to its content. Prince Cesi had specifically invited Galileo to his home outside Rome to prepare him for that audience, so Galileo could reach the Vatican 'not in the dark but well informed as to what might be necessary' in returning to the subject of Copernicus.

Even as Prince Cesi and Galileo discussed these prospects, news from Rome on 11 April dampened their spirits: Virginio Cesarini, the prince's cousin and their fellow Lyncean Academician, immortalised as the addressee of *The Assayer*, had died at twenty-seven of tuberculosis.

MOST ILLUSTRIOUS AND BELOVED LORD FATHER

WHAT GREAT HAPPINESS WAS delivered here, Sire, along with the news (via the letter that you ordered sent to Master Benedetto [Landucci]) of the safe progress of your journey as far as Acquasparta, and for all of this we offer thanks to God, Master of all. We are also delighted to learn of the favours you received from Prince Cesi, and we hope to have even greater occasion for rejoicing when we hear tell of your arrival in Rome, Sire, where persons of grand stature most eagerly await you, even though I know that your joy must be tainted with considerable sorrow, on account of the sudden death of Signor Don Virginio Cesarini, so esteemed and so loved by you. I, too, have been saddened by his passing, thinking only of the grief that you must endure, Sire, for the loss of such a dear friend, just when you stood on the verge of soon seeing him again; surely this event gives us occasion to reflect on the falsity and vanity of all hopes tied to this wretched world.

But, because I would not have you think, Sire, that I want to sermonise by letter, I will say no more, except to let you know how we fare, for I can tell you that everyone here is very well indeed, and all the nuns send you their loving regards. As for myself, I pray that our Lord grant you the fulfilment of your every just desire.

FROM SAN MATTEO, THE 26TH DAY OF APRIL 1624.
Most affectionate daughter,

S. M. Celeste

This is the only letter of Suor Maria Celeste's that Galileo salvaged from the tumult of 1624. More than a year separates it from the next in his collection, so that her response to his intercession with the pope on San Matteo's behalf is lost in the lacuna. (That Galileo petitioned effectively, however, is borne out by later letters that mention 'our Father Confessor' in complete comfort, as someone who sends regards to Galileo, even watches Galileo's house for him when he is out of town, and once asks for help settling some personal business in Rome.)

Galileo rode into Rome on 23 April. The following day, Pope Urban received him congenially in a private audience for the first time – and then five more times in as many weeks over the course of Galileo's stay. The old friends strolled through the Vatican Gardens for an hour at a time, treating all the topics Galileo had hoped to discuss with His Holiness.

Although no one recorded the content of Galileo's springtime sessions with Urban in 1624, there can be little doubt they assessed the fallout from the momentous decree that had dominated their last days together. Many Italian scientists felt their hands tied by the Edict of 1616. Outside Italy, however, few heeded the anti-Copernican ruling. As Galileo probably knew from his correspondents across Europe, no astronomer in France, Spain, Germany or England had even bothered to make the required corrections to *De revolutionibus* published in 1620. In a sense, the edict had made Italy lose face among scientists abroad. There were rumours, too – to make Urban wince – of Germans on the verge of converting to Catholicism who backed away because of the edict.

Urban, now more than halfway through the first year of his pontificate, was proud to say he had never supported that decree, and that it would not have seen the light had he been pope in those days. As a cardinal, he had successfully intervened, along with his colleague Bonifazio Cardinal Caetani, to keep 'heresy' out of the edict's final wording. Thus, although the consultors to the Holy Office had called the immobility of the Sun 'formally heretical' in their February 1616 report, the 5 March edict merely stated that the doctrine was 'false' and 'contrary to Holy Scripture'.

Engraving of Galileo at age sixty, by Ottavio Leoni

Why had Maffeo Barberini, a man with no vested interest in the Sun-centred universe, taken such action? His admiration for Galileo could well have figured in his thinking. But he no doubt had other reasons, too. Both Cardinal Barberini and Cardinal Caetani, having studied some astronomy, distinguished themselves from theologians who never looked up to Heaven except to pray. Neither cardinal believed in the physical reality of the heliocentric universe, of course, but they recognised its merit as a way of thinking about cosmology. They also valued *De revolutionibus* itself as a mathematical *tour de force*, and they wanted to preserve the intellectual freedom of Catholic scholars to read it – pending certain revisions. (Cardinal Caetani argued so strongly in favour of the book, in fact, that he was later chosen as the one to amend it.)

The eight years since the edict had not swayed Urban from his position on Copernicus. He still saw no harm in using the Copernican system as a tool for astronomical calculations and predictions. The Sun-centred universe remained merely an unproven idea – *without*, Urban felt certain, any prospect of proof in the future. Therefore, if Galileo wished to apply his science and his eloquence to a consideration of Copernican doctrine, he could proceed with the pope's blessing, so long as he labelled the system a hypothesis.

By the time Galileo started back to Florence on 8 June, he had secured not only the promise of a pension for Vincenzio and redress for San Matteo, but a personal letter from Urban to young Ferdinando, in which the pope lauded the grand duke's premier philosopher: 'We embrace with paternal love this great man whose fame shines in the heavens and goes on Earth far and wide.'

All these favourable words and gestures heartened Galileo, convinced him that he could indeed resume his public musings about an Earth in motion about the Sun. Before attempting the book-length tome, however, Galileo decided to write something along its lines as a trial run, by replying to an anti-Copernican treatise that had circulated through Rome since 1616. Though unpublished, the still uncontested comments of Monsignor Francesco Ingoli, secretary of the Congregation of the Propagation of the Faith, seemed to beg for a response – especially as these comments had originated in a debate with Galileo.

In 1616, during Galileo's aggressive Copernican campaign in Rome, he had staged one of his evening disputes against the very same Ingoli. Afterwards, the two of them agreed to write down their respective positions. No sooner had Ingoli done his half, however, than the Edict of 1616 intruded, leaving the written phase of the contest incomplete. Even now, Galileo hesitated to take on a man like Ingoli, who had based many of his points on theology instead of astronomy. However, he began drafting the ticklish response immediately upon his return to Bellosguardo.

'Eight years have already passed, Signor Ingoli,' Galileo began, 'since while in Rome I received from you an essay written almost in the form of a letter addressed to me. In it you tried to demonstrate the falsity of the Copernican hypothesis, concerning which there was much turmoil at that time.'

Having refreshed Ingoli's memory of the events, Galileo excused his own silence as the only appropriate response to the weakness of Ingoli's arguments. Galileo could have swatted these away in a single blow – except, of course, the theological ones – but he simply hadn't bothered to refute Ingoli because he deemed the effort a waste of time and breath.

'However,' Galileo continued, 'I have now discovered very tangibly that I was completely wrong in this belief of mine: having recently gone to Rome to pay my respects to His Holiness Pope Urban VIII, to whom I am tied by an old acquaintance and by many favours I have received, I found it is firmly and generally believed that I have been silent because I was convinced by your demonstrations . . . Thus I have found myself being forced to answer your essay, though, as you see, very late and against my will.'

Galileo applied all his talents of tact in this letter, not so much to suffer Ingoli – whom he variously insulted for bringing forth absurdities, for basing discussions on idle imagination and for uttering great inanities – but rather to handle the theological overtones of the Copernican controversy with all necessary caution: 'Note, Signor Ingoli, that I do not undertake this task with the thought or aim of supporting as true a proposition which has already been declared suspect and repugnant to a doctrine higher than physical and astronomical disciplines in dignity and authority.'

Instead, Galileo made it clear that he replied to rectify his own reputation, and to show Protestants to the north who had no doubt read Ingoli's manuscript – notably Kepler – that Catholics in general understood much more about astronomy than Ingoli's essay might lead them to believe.

'I am thinking about treating this topic very extensively,' confessed Galileo, 'in opposition to heretics, the most influential of whom I hear accept Copernicus's opinion; I would want to show them that we Catholics continue to be certain of the old truth taught us by the sacred authors, not for lack of scientific understanding, or for not having studied as many arguments, experiments, observations and demonstrations as they have, but rather because of the reverence we have towards the writings of our Fathers and because of our zeal in religion and faith.'

Italian astronomers, in other words, could tolerate the cognitive dissonance of admiring Copernicus on a theoretical level, while rejecting him theologically. 'Thus, when they [Protestants] see that we understand very well all their astronomical and physical reasons, and indeed also others much more powerful than those advanced till now, at most they will blame us as men who are steadfast in our beliefs, but not as blind to and ignorant of the human disciplines; and this is something which in the final analysis should not concern a true Catholic Christian – I mean that a heretic laughs at him because he gives priority to the reverence and trust which is due to the sacred authors over all the arguments and observations of astronomers and philosophers put together.'

Galileo could not repeat his real reasons for embracing both Copernicus and the Bible, as expressed in his *Letter to the Grand Duchess Cristina*, for the edict had indeed prohibited that kind of exegesis. For the moment – and the unforeseeable future – he let his Catholic faith dictate the form in which he framed his arguments.

From this safe position, Galileo found he could once again dare to defend Copernicus:

> For, Signor Ingoli, if your philosophical sincerity and my old regard for you will allow me to say so, you should in all honesty have known that Nicolaus Copernicus had spent

more years on these very difficult studies than you had spent days on them; so you should have been more careful and not let yourself be lightly persuaded that you could knock down such a man, especially with the sort of weapons you use, which are among the most common and trite objections advanced in this subject; and, though you add something new, this is no more effective than the rest. Thus, did you really think that Nicolaus Copernicus did not grasp the mysteries of the extremely shallow Sacrobosco?* That he did not understand parallax? That he had not read and understood Ptolemy and Aristotle? I am not surprised that you believed you could convince him, given that you thought so little of him. However, if you had read him with the care required to understand him properly, at least the difficulty of the subject (if nothing else) would have confused your spirit of opposition, so that you would have refrained or completely abstained from taking such a step.

Since what is done is done, let us try, as far as possible, to prevent you or others from multiplying errors. So I come to the arguments you give to prove that the Earth, and not the Sun, is located at the centre of the universe.

Galileo's ire rises through the response, aggravated by his frustration with Ingoli's faulty logic, leading him to an emphatic statement of his own position: 'If any place in the world is to be called its centre, that is the centre of celestial revolutions; and everyone who is competent in this subject knows that it is the Sun rather than the Earth which is found therein.'

Galileo sent off his fifty-page 'Reply to Ingoli' to friends in Rome in October of 1624. Curiously, because of lengthy delays caused by changes Prince Cesi and other Roman colleagues wished to insert for prudence's sake, the 'Reply to Ingoli' never reached Ingoli himself. A few manuscript copies circulated cautiously around Rome, however, and the pope was treated to at

---

* This was the Latinised name of thirteenth-century English astronomer John of Holywood, who wrote the influential textbook *Sphere of Sacrobosco*.

least a partial private reading in December. No explosion erupted from Urban in reaction to the 'Reply'. Indeed, His Holiness remarked on the aptness of its examples and experiments. And therefore no apparent obstacle stood in the way of Galileo's expressing the same ideas in a book, which he now envisioned as a playlike discussion among a group of fictional friends, with the working title 'Dialogue on the Tides'.

[XIII]

Through my memory of their eloquence

GALILEO THREW HIMSELF INTO the work on the new book with all the force that his science, his religion, his life experience and his flair for the dramatic allowed. The topic deserved nothing less.

'The constitution of the universe', Galileo said in dedicating his *Dialogue* to the grand duke of Tuscany, 'may be set in first place among all natural things that can be known. For coming before all others in grandeur by reason of its universal content, it must also stand above them all in nobility as their rule and standard. Therefore if any men might claim extreme distinction in intellect above all mankind, Ptolemy and Copernicus were such men, whose gaze was thus raised on high and who philosophised about the constitution of the world. These dialogues of mine . . . set forth the teaching of these two men whom I consider the greatest minds ever to have left us such contemplations in their works.'

The writing of the *Dialogue* occupied him intermittently for a period of six years following his 1624 meetings with Pope Urban. Since Galileo had actually begun thinking about the subject matter long before that, however, and even previously written the backbones of certain sections in non-dialogue form, such as his 'Treatise on the Tides', he personally considered the book the product of a decade's gestation.

Galileo had experimented with the dialogue style in his humorous exposition on the nova of 1604, as well as in actual playwriting for his family and friends, and his father had employed dialogue to write about ancient and modern music. Aside from the dialogue's popularity for presenting scientific issues, the format offered Galileo a measure of protection: by putting the shortcomings of Ptolemy – and the merits of Copernicus – into the mouths of dramatis personae, the author could distance himself from their sensitive discussions as though he were an impartial bystander. In addition, the dialogue form allowed the characters occasionally to veer off the main theme to treat other topics – magnetism,* for example – that Galileo deemed 'no less interesting than the principal argument'.

His book took the form of an animated encounter, spread over four days, like a play in four acts, among three acquaintances who breathed their own personalities into the theories they entertained. The character he called Salviati, a thinly disguised alter ego, spoke Galileo's own mind; Sagredo, an intelligent and receptive man of means, typically took Salviati's side. Simplicio, on the other hand, a pompous Aristotelian philosopher who loved to drop Latin phrases, would often wax prolix on a topic before being played for a fool. Galileo also inserted himself as a minor character in the work by having the men cite the authority of 'the Lyncean Aca-

---

* Galileo read and admired the 1600 opus *De magnete* by English scientist William Gilbert (1544–1603).

demician' from time to time, or allude to the discoveries and ideas of 'our mutual friend'.

Galileo, over sixty now, had reached the stage in life at which many of his own dearest friends had already died. But in the *Dialogue*, he brought two of the departed back to life for all time: Filippo Salviati, the gracious host at whose Florentine Villa delle Selve Galileo had resided for long periods, writing while recuperating from illnesses, and Giovanfrancesco Sagredo, who had been a private student of Galileo's at Padua and kept in close touch with him till his death in Venice in 1620. The name Simplicio recalled no particular colleague of Galileo's, but rather the sixth-century Greek philosopher Simplicius, a renowned commentator on Aristotle. Behind that ancient identity hid some unspecified pedant – thought to be Cesare Cremonini, University of Padua philosopher – who had frequently opposed Galileo in debate. Of course, the name Simplicio also sounded something like *sempliciotto*, or *simpleton*, which could hardly have escaped Galileo's notice and might well have been his intention.

'Many years ago in the marvellous city of Venice,' Galileo recounted in the *Dialogue*'s preface, 'I had several occasions to engage in conversation with Giovanfrancesco Sagredo, a man of most illustrious family and of sharpest mind. From Florence we were visited by Filippo Salviati, whose least glory was purity of blood and magnificence of riches; his sublime intellect fed on no delight more avidly than on refined speculations.' Galileo had often drawn these two out on the three main topics to be considered in the *Dialogue*, namely, the question of the Earth's motion, the organisation of the heavenly bodies, and the ebb and flow of the sea.

'Now, since bitter death has deprived Venice and Florence of those two great luminaries in the very meridian of their years, I have decided to prolong their existence, so far as my poor abilities will permit, by reviving them in these pages of mine and using them as interlocutors in the present controversy . . . May it please those two great souls, ever venerable in my heart, to receive with favour this public monument of my undying friendship. And may they assist me, through my memory of their eloquence, to explain to posterity the promised reflections.'

The action of the *Dialogue* unfolds at Sagredo's palazzo in Venice, where his guests, Salviati and Simplicio, arrive each day by gondola. The three have earmarked these four days, according to the conceit of the book, for their mutual enlightenment – to isolate themselves for an intellectual retreat so as to 'discuss as clearly and in as much detail as possible' the two chief systems of the world (as the universe was then commonly called).

Writing in Italian for a mass audience, Galileo laced the five-hundred-page *Dialogue* with grand, gorgeous language, by turns poetic, didactic, reverent, combative and funny. He illustrated the text, too, but only sparsely, by having his characters create simple line drawings for each other as need arose. 'Just seeing the diagram has cleared the whole matter up,' Sagredo says gratefully to the more learned Salviati at one such juncture, 'so go on.'

In the margins of almost every page, where Galileo normally penned notes to himself in the books he read, he put postils – short phrases that described the content of the paragraphs or reiterated their main points. A few densely numerical tables of observational data also entered the *Dialogue*, but not until the morning of Day Three, by which time the lay reader had been prepared either to absorb them or skip over them without losing the gist.

As for the central dilemma of the *Dialogue* – the challenge of championing Copernicus without alienating the Church – Galileo faced it on the very first page of his preface. Here he took pains to explain the delicate situation in Italy, where a scientist (he, Galileo) had made vital discoveries pertaining to the Copernican doctrine, but where the doctrine itself ran foul of the reigning religious authorities. He had intimated as much in his 'Reply to Ingoli'. Now he stated his position for the record, addressing 'the discerning reader' as follows: 'Some years ago there was published in Rome a salutary edict which, in order to obviate the dangerous tendencies of our present age, imposed a seasonable silence upon the Pythagorean opinion that the Earth moves. There were those who impudently asserted that this decree had its origin not in judicious inquiry, but in passion none too well informed. Complaints were to be heard that advisers who were totally unskilled at astronomical observations ought not to clip the wings of reflective intellects by means of rash prohibitions.' Indeed, Galileo himself had expressed exactly these sentiments in his *Letter to the Grand Duchess Cristina*. But he wrote that plea before the edict was issued. After the fact of the edict, he did not gainsay the Holy Fathers. The *Dialogue* resumed his importuning that truths about Nature be allowed to emerge through science. Such truths, he still believed, could only glorify the Word and deeds of God.

'Upon hearing such carping insolence', he continued in the preface,

> my zeal could not be contained. Being thoroughly informed about that prudent determination, I decided to appear openly in the theatre of the world as a witness of the sober truth. I was at that time in Rome; I was not only received by the most eminent prelates of that Court, but had their applause; indeed, this decree was not published without some previous notice of it having been given to me. Therefore I propose in the present work to show to foreign nations that as much is understood of this matter in Italy, and particularly in Rome, as trans-Alpine diligence can ever have imagined. Collecting all the reflections that properly concern the Copernican system, I shall make it known that everything was brought before the attention of the Roman censorship, and that there proceed from this clime not only dogmas for the welfare of the soul, but ingenious discoveries for the delight of the mind as well.

Then Day One opens like a curtain rising, with the characters already assembled and their conversation coming immediately to the point. The day's discussions draw the dividing lines between the Aristotelian/Ptolemaic and the Copernican world-views. To that end, Simplicio propounds Aristotle's view of the Earth as fundamentally different from all celestial bodies – in that it comprises elements instead of ether. Salviati, *à la* Galileo, seeks to find the Earth a place in Heaven. And Sagredo good-naturedly grants the Earth – 'the dregs of the universe, the sink of all uncleanness' – a unique strength born of its characteristic susceptibility to change: 'For my part I consider the Earth very noble and admirable precisely because of the diverse alterations, changes, generations, etc. that occur in it incessantly. If, not being subject to any changes, it were a vast desert of sand or a mountain of jasper, or if at the time of the flood the waters which covered it had frozen, and it had remained an enormous globe of ice where nothing was ever born or ever altered or changed, I should deem it a useless lump in the universe, devoid of activity and, in a word, superfluous and essentially nonexistent.'

Salviati backs him up with the evidence from their 'friend's' telescope about how the Sun also changes, via spots erupting and receding round and round its girth. He suggests that the Moon, too, may be changing, and all the stars as well, fixed or wandering, but that the changes have thus far escaped detection. Immutability, which Aristotle had deemed a defining quality of perfect orbs, here dissolves into a simple lack of information.

'The deeper I go in considering the vanities of popular reasoning,' the host Sagredo resumes his musing,

> the lighter and more foolish I find them. What greater stupidity can be imagined than that of calling jewels, silver and gold 'precious', and earth and soil 'base'? People who do this ought to remember that if there were as great a scarcity of soil as of jewels or precious metals, there would not be a prince who would not spend a bushel of diamonds and rubies and a cartload of gold just to have enough earth to plant a jasmine in a little pot, or to sow an orange seed and watch it sprout, grow, and produce its handsome leaves, its fragrant flowers and fine fruit. It is scarcity and plenty that make the vulgar take things to be precious or worthless; they call a diamond very beautiful because it is like pure water, and then would not exchange one for ten barrels of water. Those who so greatly exalt incorruptibility, inalterability, etc. are reduced

to talking this way, I believe, by their great desire to go on living, and by the terror they have of death. These individuals do not reflect that if men were immortal, they themselves would never have come into the world. Such men really deserve to encounter a Medusa's head which would transmute them into statues of jasper or of diamond, and thus make them more perfect than they are.

Illness interrupted Galileo's progress on the *Dialogue* in March of 1625. Though he recovered relatively quickly, he did not return to the book right away. Guiducci, his former live-in student and co-author in the comet controversy, wrote to him from Rome that some unnamed 'pious person' had lodged complaints with the Holy Office of the Inquisition about *The Assayer*, on the grounds that it undermined the Sacrament of the Eucharist. Galileo had commented in *The Assayer* on the nature of matter – how it breaks down into even smaller parts that lose all resemblance to familiar objects. This philosophy seemed to question the integrity of the bread offered in the mass as the body of Christ, and the wine that was His blood. As a precaution, Guiducci advised Galileo to muffle his 'Reply to Ingoli' for the time being, since its praise of Copernicus was unabashed. And Galileo took the further prudent measure of suspending work on the *Dialogue*.

Branch of an orange tree

He occupied himself with other interests of his own and his official duties as philosopher to the grand duke, who called on Galileo to evaluate the schemes and machines that enterprising inventors tried to sell the Tuscan government. One such offering purported to pump water with miraculous efficiency, and another to revolutionise the milling of grain. After attending model demonstrations, Galileo would write polite but often daunting letters to the inventors, explaining the hopelessness of their ideas on the basis of the principles of simple machines.

'I cannot deny that I was admiring and confused when, in the presence of the Grand Duke and other princes and gentlemen, you exhibited the model of your machine, of truly subtle invention,' Galileo began his critique of the would-be water pump. 'And since I long ago formed the idea, confirmed by many experiments, that Nature cannot be overcome and defrauded by art,' he added soon after, 'I have made an accumulation of thoughts and have decided to put them on paper and communicate them to you in order that if the success of your truly acute invention is seen in practice, and in very large machines, I may be excused by you, and through you excused by others.'

At the same juncture in 1625, Galileo also gave mathematical critiques on papers he received from his correspondents in Pisa, Milan, Genoa, Rome and Bologna concerning their thoughts on the dynamics of river flow, the refraction of light, the acceleration of bodies in fall and the nature of indivisible points.

In his spare time Galileo repaired to his garden, where he indulged the pleasure he had described of planting orange seeds – and lemons and chartreuse-coloured citrons – in large terracotta pots. Galileo regularly sent the best of the citrons to Suor Maria Celeste, who would seed, soak, dry and sweeten them over a period of several days to prepare his favourite confection. Having fared poorly with the fruits he consigned to her just before Christmas 1625, however, she sought a few other tokens she hoped would please him.

Most Illustrious and Beloved Lord Father

As for the citron, which you commanded me, Sire, to make into a sweetmeat, I have come up with only this little bit that I send you now, because I am afraid the fruit was not fresh enough for the confection to reach the state of perfection I would have liked, and indeed it did not turn out very well after all. Along with this I am sending you two baked pears for these festive days. But to present you with an even more special gift, I enclose a rose, which, as an extraordinary thing in this cold season, must be warmly welcomed by you. And all the more so since, together with the rose, you will be able to accept the thorns that represent the bitter suffering of our Lord; and also its green leaves, symbolising the hope that we nurture (by virtue of this holy passion), of the reward that awaits us, after the brevity and darkness of the winter of the present life, when at last we will enter the clarity and happiness of the eternal spring of Heaven, which blessed God grants us by His mercy.

And ending here, I give you loving greetings, together with Suor Arcangela, and remind you, Sire, that both of us are all eagerness to hear the current state of your health.

From San Matteo, the 19th day of December 1625.
Most affectionate daughter,

S. M. Celeste

I am returning the tablecloth in which you wrapped the lamb you sent; and you, Sire, have a pillowcase of ours, which we put over the shirts in the basket with the lid.

The garden at the Convent of San Matteo in Arcetri, where a rose could bloom at Christmastime, devoted much of its earthly paradise to herbs and medicinal plants such as rosemary (good for treating nausea) and rue (applied directly to the nostrils to stanch a bloody nose, or drunk with wine for headache). These grew among the pine, plum and pear trees ranged around a central well at the back of the church. Even the decorative rosebushes served an apothecary's purpose, for the syrup of cooked, compressed rosebuds made an excellent purgative (prepared from several hundred roses, picked when the buds were half open, then steeped a full day and night in sugar and hot water). Just beyond the garden, the fragrant almond trees and evergreen olive trees cascaded down the slope behind the convent, where walkways led the nuns easily into Franciscan communion with Nature.

The convent's walled perimeter, not to mention the rule of enclosure, detained Suor Maria Celeste in an anteroom of the afterlife. Rather than resent this separation from worldly affairs, cloistered monks and nuns at that time typically developed fierce attachments to their self-contained communities, where some spent long lifetimes (like the aunt of Pope Urban VIII who lived eighty-one years at her convent), and where, according to contemporary record books, miracles occurred as matters of course. A statue of the Blessed Virgin might weep or bow her head above the blue-flowered rosemary shrubs. Bones of saints buried in the on-site cemetery could be heard rattling to herald the death of a nun.

Florentine monasteries also guarded an abundance of holy relics, including fifty-one authentic thorns from the crown of Jesus and the tunic worn by Saint Francis of Assisi when the stigmata first appeared on his body.

The difference Suor Maria Celeste discerned between this vale of tears and the harmony of Paradise precisely echoed Aristotle's distinction between corruptible Earthly matter and the immutable perfection of the heavens. This consonance was no coincidence, but the fruit of the labours of the prolific Italian theologian Saint Thomas Aquinas, who grafted the third-century-BC writings of Aristotle on to thirteenth-century Christian doctrine. The compelling works of Saint Thomas Aquinas had reverberated through the Church and the nascent universities of Europe for hundreds of years, helping the word of Aristotle gain the authority of Holy Writ, long before Galileo began his book about the architecture of the heavens.

[XIV]

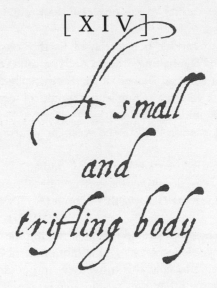

A small
and
trifling body

BY 1626, GALILEO HAD neglected his *Dialogue* for so long that his friends feared he might never return to it. And if not Galileo, then who would step forward to correct humanity's self-centred view of the cosmos? Who better than Galileo to propound the most stunning reversal in perception ever to have jarred intelligent thought? We are not the centre of the universe. The immobility of our world is an illusion. We spin. We speed through space. We circle the Sun. We live on a wandering star.

The apparent steadiness of the Earth lulls the mind into a false stability. The body's footing feels so secure that the mind naturally interprets the daily bobbing up and down of the Sun, the Moon, the planets and the stars as motions entirely external to the Earth. Even at night, under the open sky, assaulted by the intimations of infinity scintillating through the cope of Heaven, the mind would rather cede revolution to the universe than relinquish the solace of solid ground.

This incontrovertible perception of Earthly rest gains support on every hand. The halting drop of each autumn leaf adds weight to the case for stillness. Indeed, if the Earth really turned towards the east at high velocity, falling leaves would all scatter to the west of the trees. Wouldn't they?

Wouldn't a cannon fired to the west carry further than a salvo to the east?

Wouldn't birds lose their bearings in mid-air?

These questions doubting the Earth's diurnal motion consumed the participants in the *Dialogue* throughout Day Two of their deliberations.

Here Galileo demonstrated that the moving Earth – if it did indeed move – would impart its global motion to all Earthly objects. Instead of having to fight this movement in one direction and be abetted by it in the other, they would own it themselves, just as the passengers aboard a ship may saunter up or down the decks as their bodies travel – with more rapid motion, through no effort of their own – the two thousand miles from Venice to Aleppo.

As he built up the evidence for the Copernican thesis, Galileo interjected protective reminders regarding his own neutrality. 'I act the part of Copernicus in our arguments and wear his mask,' Salviati explains to his two companions. 'As to the internal effects upon me of the arguments which I produce in his favour, I want you to be guided not by what I say when we are in the heat of acting out our play, but after I have put off the costume, for perhaps then you shall find me different from what you saw of me on the stage.'

Thus freed to debate all the more forcefully, Salviati shows how a cannon pointed east moves east along with the Earth, as does the cannonball loaded inside it. After firing, therefore, the eastwards shot has no trouble flying as far as a westwards one. Nor do birds, though they fly slower than the Earth turns, find their flight adversely affected by the diurnal rotation. 'The air itself through which they wander,' notes Salviati, 'following naturally the whirling of the Earth, takes along the birds and everything else that is suspended in it, just as it carries the clouds. So the birds do not have to worry about following the Earth, and so far as that is concerned they could remain for ever asleep.'

As participants in the Earth's activity, people cannot observe their own rotation, which is so deeply embedded in terrestrial existence as to have become insensible.

'Shut yourself up with some friend in the main cabin below decks on some large ship,' Salviati suggests later on the second day,

> and have with you there some flies, butterflies and other small flying animals. Have a large bowl of water with some fish in it; hang up a bottle that empties drop by drop into a wide vessel beneath it. With the ship standing still, observe carefully how the little animals fly with equal speed to all sides of the cabin. The fish swim indifferently in all directions; the drops fall into the vessel beneath; and, in throwing something to your friend, you need throw it no more strongly in one direction than another, the distances being equal; jumping with your feet together, you pass equal spaces in every direction. When you have observed all these things carefully (though there is no doubt that when the ship is standing still everything must happen in this way), have the ship proceed with any speed you like, so long as the motion is uniform and not fluctuating this way and that. You will discover not the least change in all the effects named, nor could you tell from any of them whether the ship was moving or standing still.

Remarkably, Galileo conceded through this early demonstration in relativity, *no experiment* performed with ordinary objects on the surface of the Earth could prove whether or not the world was actually turning. Only astronomical evidence and reasoning from simplicity could carry the argument. Thus the daily rotation of the Earth streamlined the hubbub of the universe and recognised the cosmic balance of power. For if the heavens really revolved with enough force to propel the vast bodies of the innumerable stars, how could the puny Earth resist the tide of all that turning?

'We encounter no such objections', Salviati replies to his own rhetorical question, 'if we give the motion to the Earth, a small and trifling body in comparison with the universe, and hence unable to do it any violence.'

When evening falls on the long second day of the *Dialogue*, and the three friends exchange their goodbyes, Simplicio leaves promising to review the modern Aristotelian arguments against the Earth's annual motion, in preparation for the next day's discussion. Salviati, too, stays up late to study an anti-Copernican text about comets and novas that Simplicio has lent him, and Sagredo lies awake, his thoughts flitting from one cosmological order to the other, eager for a resolution. But somewhere in that second day's night of reflection, Galileo stopped writing, so that the conversation among the *Dialogue*'s three characters hung suspended for several years while their author continued thinking through the intricate proofs to be presented, and also tended to other pressing affairs.

The long-promised pension for Galileo's son at last became real in 1627 through a bull issued by Pope Urban VIII. Vincenzio was to receive a canonry up north at Brescia, along with an annual income of sixty *scudi* – except that he refused the offer. Still attending school at Pisa, twenty-one-year-old Vincenzio admitted that he felt a 'bitter hatred of the clerical state'. This confession drove a new wedge between him and his father, and also compelled Galileo to find another recipient for the pension itself. A candidate soon emerged from the ranks of his extended family.

In May of 1627, Galileo received a letter from his brother, Michelangelo, who now carried on their father's unremunerative musical profession in Munich. Michelangelo proposed to send his wife, Anna Chiara, along with a few of their eight children, to stay with Galileo in Florence for an unspecified length of time, where the family would be safe from the German turmoil of what eventually came to be called the Thirty Years' War. This conflict had begun in 1618, when enraged Protestant noblemen threw two Catholic governors out of the windows of the royal castle in Prague. The 'Defenestration of Prague' renewed long-dormant religious bloodshed between Protestant and Roman

Papal seal of Urban VIII

Catholic states in Germany. In no time the conflict drew in neighbouring countries warring for the political prize at stake: control of the Holy Roman Empire of the German nation. By 1627, the fighting, almost entirely confined to German soil, embroiled Dutch, Spanish, English, Polish and Danish armies. If impoverished Michelangelo needed a good excuse to export his family to Italy, the war certainly added substance to his personal reasons.

'This arrangement would be good for both of us,' Michelangelo wheedled. 'Your house will be well and faithfully governed, and I should be partly lightened of an expense which I do not know how to meet; for Chiara would take some of the children with her, who would be an amusement for you and a comfort to her. I do not suppose that you would feel the expense of one or two mouths more. At any rate, they will not cost you more than those you have about you now, who are not so near akin, and probably not so much in need of help as I am.'

Anna Chiara arrived in July, with all her children in tow, not to mention their German nurse, for a total of ten additional mouths to feed. Thus Galileo, at the age of sixty-three, suddenly found himself the head of a noisy household that included his infant niece, Maria Fulvia, two-year-old Anna Maria, Elisabetta, three unruly boys – Alberto, Cosimo and little Michelangelo – and Mechilde, the dutiful older daughter. Galileo arranged for his eldest visiting nephew, the twenty-year-old Vincenzio, to study music in Rome. He further accommodated the youth by granting to him the papal pension he had secured for his own Vincenzio.

Galileo's house at Bellosguardo, where he lived from 1617 to 1631

Although Michelangelo's Vincenzio evinced genuine musical talent, he disdained to work at his lessons. The Benedictine monk Benedetto Castelli, who had taken Galileo's son in hand in Pisa, had recently been transferred to Rome, where he now tried vainly to keep Galileo's nephew in line. But this Vincenzio stayed out all night with rapscallions, ran up debts, and flouted religious decorum so flagrantly that his Roman landlord threatened to have him denounced to the Holy Office of the Inquisition.

In Brescia, meanwhile, resident clerics ignored Galileo's transfer of the pension from one Vincenzio Galilei to another: they elected a citizen of their own town to serve as canon. (Galileo waited a few years for the popular Brescian to die and create a new vacancy before he tried to install another family member in that post.)

With his house filled to the rafters, Galileo abandoned Bellosguardo when he next fell ill in mid-March of 1628, taking refuge at the home of acquaintances in Florence.

'Something in the peaceful air today', Suor Maria Celeste wrote to him when he had recovered and returned to his own quarters, gave me half a hope of seeing you again, Sire. Since you did not come, we were most pleased with the arrival of adorable little Albertino, along with our Aunt, giving us news that you are well and that you will soon be here to see us; yet my delight was all but destroyed when I learned that you have already returned to your usual labours in the garden, leaving me considerably disturbed; since the air is still quite raw, and you, Sire, still weak from your recent illness, I fear this activity will do you harm. Please do not forget so quickly the grave condition you were in, and have a little more love for yourself than for the garden; although I suppose it is not for love of the garden per se, but rather the joy you draw from it, that you put yourself at such risk. But in this season of Lent, when one is expected to make certain sacrifices, make this one, Sire: deprive yourself for a time of the pleasure of the garden.

Galileo had considered undertaking a second pilgrimage to Loreto in 1628, marking the decade since his last call at the Casa Santa. He mentioned the possibility in a letter to his brother, and even suggested taking his sister-in-law, Anna Chiara, along with him on the long trek, but his illness prevented him from making such a trip. Instead, Anna Chiara and her children, whom Suor Maria Celeste now referred to as 'the little rabble-rousers', took their leave. Catholic forces in Germany seemed virtually assured of victory at this point, and in any case, after nearly a year, it was time to go home.

Soon after his long-visiting relatives returned to Germany in the spring of 1628, Galileo fell into a tenure wrangle with the University of Pisa. His original appointment to the Tuscan court in 1610 had included a lifetime chair on the university faculty, conferred by Cosimo II. Galileo occupied the honorary position *in absentia*, with no obligation even to visit Pisa, and no mandatory teaching duties to derail him from his more important mission of discovering novelties and announcing them to the world at large for the greater glory of the grand duke. But the university administration, tired of honouring the old contract, suddenly moved to dissolve it. Galileo fought back with vigour, and with the full support of Ferdinando II, for although his professorial appointment had come through the court, the university paid Galileo's salary of one thousand *scudi* annually.

By April, Galileo told Suor Maria Celeste of his plans to go to Pisa. In addition to the legal proceedings with the university (which wore on, with and without his presence, for nearly two years before Galileo emerged victorious), he meant to attend Vincenzio's graduation. Suor Maria Celeste pressed him to do a favour *en route* for two poor nuns at the convent by purchasing several yards of cheap Pisan wool cloth, for which they had pooled together eight *scudi*.

At the Pisan academic ceremonies, Vincenzio collected his doctoral degree in law as the culmination of six years of study. (Galileo himself had never experienced this rite of passage, having left school without completing his required courses.) A more mature Vincenzio now returned home to his father's house at Bellosguardo, where he turned twenty-two in August. He took to paying regular visits to his sisters at San Matteo. And the following December he delighted Galileo with his decision to marry young Sestilia Bocchineri, whose family held prestigious appointments in the Tuscan court.

MOST BELOVED LORD FATHER

THE UNEXPECTED NEWS DELIVERED here by our Vincenzio regarding the finalisation of his wedding plans, and marrying into that esteemed family, has brought me such happiness that I would not know how better to express it, save to say that, as great as is the love I bear you, Sire, equally great is the delight I derive from your every joy, which I imagine in this instance to be immense; and therefore I come now to rejoice with you, and pray our Lord to protect you for a very long time, so that you can savour those satisfactions that seem guaranteed to you by the good qualities of your son and my brother, for whom my affection grows stronger every day, as he appears to me to be a calm and wise young man.

I would much rather have celebrated with you in person, Sire, but if you would be so kind, I implore you to at least tell me by letter how you plan to arrange your visit with Vincenzio's betrothed: meaning whether it may be well to meet in Prato when Vincenzio goes there, or better to wait until she is in Florence, since this is the usual formality among us sisters, and surely, given her experience of having been in a convent, she will know these customs. I await your resolution. And in the meanwhile I bid you adieu from my heart.

Your most affectionate daughter,

S. M. Celeste

[XV]

On the right path, by the grace of God

THE FAMILY OF SESTILIA Bocchineri, Vincenzio's intended, lived above reproach about twelve miles from Florence in the textile city of Prato, famed for its mulberry trees and silkworms. There her father served as major-domo at the Medici's local palace, while her brother Geri held a Florentine court position at the Pitti. The Tuscan secretary of state had contrived to couple Vincenzio with Sestilia, entirely to Galileo's satisfaction. The bride brought a dowry of seven hundred *scudi* to the marriage – an amount that would have secured her future in her convent had she not been betrothed instead on the eve of her sixteenth birthday. The wedding date, set for 29 January 1629, allowed the relatives about a month's time for preparations.

Although Suor Maria Celeste could not leave San Matteo to attend any of her brother's nuptial festivities, she determined to cater at least part of the wedding feast, to spare her father yet another financial burden. 'Here then is a list of the more costly items that we will need for making a platter of pastries,' she wrote to him on 4 January, enclosing a short shopping list of pricey staples – white sugar, almonds and fine confectioners' sugar – on a small separate square of paper, 'leaving the less expensive ingredients to me. After this, Sire, you will be able to see if you want me to cook other sweet things for you, such as savouries and the like; because I firmly believe you would spend less this way than buying them already prepared by the grocer's shop, and we will apply ourselves to making them with the utmost possible care.'

As for the bride's gift, 'My thoughts lean towards sewing her a beautiful apron', Suor Maria Celeste said later in the same letter, 'so as to give her something that would be useful, and not require a great expenditure on our part, since we could do all the work ourselves; not to mention that we have no idea how to fashion the high collars and ruffs that ladies are wearing nowadays.

'I might think I had blundered, Sire,' she added, 'asking your opinion on these many trifles, if I were not absolutely certain that you, in small details just as in great matters, know far more than we nuns do.'

For his own gift, Galileo helped the newly-weds buy a putty-coloured house with a garden and courtyard on the Costa San Giorgio, part-way up the road that ascended from Florence to Arcetri.

The month after the wedding, in the course of Galileo's continuing correspondence with his faithful friend Castelli, the monk raised the old subject of the solar spots. And Galileo, who prided himself on having long since discovered and published everything worth saying about sunspots, presently began to see a new significance in their behaviour that could confirm the Copernican world system. Within a few months, the sunspots would light Galileo's way back to his long-neglected *Dialogue*.

Castelli, aged fifty, now lived in Rome, where Pope Urban VIII had summoned him as a specialist in hydraulics to supervise water and drainage projects around the Holy City. Urban had per-

Father Christopher Scheiner

formed the same kind of consulting work himself years before, when he regulated the waters of Lake Trasimeno for Pope Clement VIII, so he recognised Castelli's expertise. While in Rome, Castelli also taught physics at the college called the Sapienza, and wrote a book in 1628 about how to measure running water. He sent Galileo a copy for comment, and during this exchange, late in February 1629, Castelli mentioned the imminent publication of a major new text on sunspots by Galileo's old rival, Father Christopher Scheiner – the 'Apelles' of the *Sunspot Letters*. (Printing problems, however, delayed the appearance of Scheiner's book, the *Rosa Ursina*, for another two years, until April 1631.)

Castelli also detailed recent observations of a gigantic sunspot that called attention to itself over a period of several weeks. The spot had crossed the body of the Sun, disappearing at length around its western limb on 9 February, only to reappear a fortnight later – still recognisable – on 24 February, on the Sun's eastern horizon.

The thought of Scheiner's reprise of sunspots must have nettled Galileo, for he denounced the forthcoming book in his April correspondence, predicting an assortment of errors and irrelevancies. In private, however, he perused his old sunspot files to see what, if anything, he might have missed before. And indeed he had missed something: he had overlooked the odd way the spots traversed the

Sun over the course of the year. Their path appeared occasionally to cut straight across the Sun's middle, and at other times to trace an upwards or a downwards slanting arc. Galileo surmised, towards the summer of 1629, that the spots probably stayed a steady course around the Sun's equator all the time. They merely *appeared* to trek uphill or down as the seasons changed because of the annual revolution of the tilted Earth around the also-tilted Sun.

The Sun thus offered its own physical evidence in support of Copernicus, compounding the testimony of the tides.

What a blow to think that Scheiner, who had foolishly mistaken sunspots for stars before Galileo corrected him, now stood ready to publish this monumental discovery! The shock impelled Galileo back to his unfinished manuscript. If he needed further incentive, he could look to the projected income from sales of the finished book, for his new daughter-in-law was already pregnant, his son still unemployed, and Suor Maria Celeste anxious to improve the quality of her life through the purchase of private quarters in the convent.

M OST BELOVED LORD FATHER

THE DISCOMFORT I HAVE endured ever since I came to live in this house, for want of a cell of my own, I know that you know, Sire, at least in part, and now I shall more clearly explain it to you, telling you that two or three years ago I was compelled by necessity to leave the one small cell we had, for which we paid our novice mistress (according to the custom we nuns observe) thirty-six *scudi*, and give it over totally to Suor Arcangela, so that (as much as possible) she could distance herself from this same mistress, who, tormented to distraction by her habitual moods, posed a threat, I feared, to Suor Arcangela, who often finds interaction with others

unbearable; beyond that, Suor Arcangela's nature being very
different from mine and rather eccentric, it pays for me to
acquiesce to her in many things, in order to be able to live in
the kind of peace and unity befitting the intense love we bear
each other. As a result I spend every night in the disturbing
company of the mistress (although I get through the nights
easily enough with the help of the Lord, who suffers me to
undergo these tribulations undoubtedly for my own good) and
I pass the days practically a pilgrim, having no place
whatsoever where I can retreat for one hour on my own. I do
not yearn for large or very beautiful quarters, but only for a
little bit of space, exactly like the tiny room that has just
become available, now that a nun who desperately needs
money wants to sell it; and, thanks to Suor Luisa's having
spoken well on my behalf, this nun prefers me over any of the
others offering to buy it. But because its price is 35 *scudi*, while
I have only ten, which Suor Luisa kindly gave me, plus the
five I expect from my income, I cannot take possession of the
room, and I rather fear I may lose it, Sire, if you do not assist
me with the remaining amount, which is 20 *scudi*.

I explain this need to you, Sire, with a daughter's security and
without ceremony, so as not to offend that loving tenderness I
have experienced so often. I will only repeat that this is of the
greatest necessity, on account of my having been reduced to
the state in which I find myself, and because, loving me the
way that I know you love me, and desiring my happiness, you
can well imagine how this step will bring me the greatest
satisfaction and pleasure, of a proper and honest sort, as all I
seek is a little quiet and solitude. You might tell me, Sire, that
to make up the sum I require, I could avail myself of the 30
scudi of yours that the convent is still holding: to which I
respond (aside from the fact that I could not lay claim to that
money quickly enough in this extreme case, as the nun selling

the room faces dire straits) that you promised the Mother Abbess you would not ask her for those funds until such time as the convent enjoyed some relief from the constraint of constant expenditures; given all that, I do not think you will forsake me, Sire, in doing me this great charitable service, which I beg of you for the love of God, numbering myself now among the neediest paupers locked in prison, and not only needy, I say, but also ashamed, since I would not dare to speak so openly of my distress to your face: no less to Vincenzio; but only by resorting to this letter, Sire, can I appeal with every confidence, knowing that you will want and be able to help me. And here to close I send you regards with all my love, and also to Vincenzio and his bride. May the Lord bless you and keep you happy always.

FROM SAN MATTEO, THE 8TH DAY OF JULY 1629.
Most affectionate daughter,

S. M. Celeste

The novice mistress, from whom Suor Maria Celeste and Suor Arcangela purchased their first small room for thirty-six *scudi*, had been chosen by the abbess as the one to 'form them in holy living and becoming behaviour', according to the practice of the order. But the mistress, who is never mentioned by name in any of Suor Maria Celeste's letters, herself abused the Rule because of serious emotional disturbances that provoked her to constant prattling.

> *The Sisters must keep silence from the hour of Compline until Terce ... Let them also be silent continually in the church, in the dormitory.*
>
> THE RULE OF SAINT CLARE, chapter V

Most of the residents of San Matteo slept in a common dormitory, though the convent contained numerous private chambers that could be bought for a price – over and above the dowry a nun's family members paid to the monastery, and in addition to the allowance they meted out to her for living expenses. In this sense, although every Poor Clare lived in poverty, some lived more poorly than others. And yet, because the sisters dined together and shared the same food, the convent's income from the sale of private quarters served to improve the fare for all.

> *It may not be allowed any Sister to send out letters, or to receive anything, or to give anything outside the monastery without permission of the Abbess. Nor is it allowed to have anything which the Abbess has not given or permitted. But if anything should be sent to a Sister by her relatives or by others, the Abbess should have it given to the Sister. Then, if she needs it, she may use it; otherwise, let her charitably give it over to a Sister in need. If, however, any money would be sent to a Sister, the Abbess with the advice of the Discreets should have provision made for her in those things she may need.*
>
> THE RULE OF SAINT CLARE, chapter VIII

Galileo sent Suor Maria Celeste the twenty *scudi*, naturally, but her living situation held even more complexity and danger than the self-conscious, convoluted language of her request could indicate. In fact she required several months to resolve her plight and find the courage to explain the full chain of events involving the money, the mistress and the cell to her father, who had resumed his ruminations on the two chief systems of the world.

'And to give you some news of my studies,' he wrote on 29 October to his lawyer friend Elia Diodati in Paris, 'you must know that a month ago I took up again my *Dialogue* about the tides, put aside for three years on end, and by the grace of God have got on the right path, so that if I can keep on this winter I hope to bring the work to an end and immediately publish it.' The two men

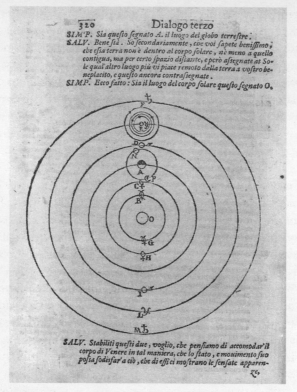

Diagram from Day Three of Galileo's *Dialogue* demonstrating
the Copernican system

had met when Diodati visited Florence in 1620 and had maintained an intellectual correspondence ever since. To Diodati, who had been born in Italy but moved first to Geneva and then to Paris because of his family's Protestant faith, Galileo could be perfectly blunt about his book's pro-Copernican slant. 'In this, besides the material on the tides, there will be inserted many other problems and a most ample confirmation of the Copernican system by showing the nullity of all that had been brought by Tycho and others to the contrary. The work will be quite large and full of many novelties, which by reason of the freedom of dialogue I shall have scope to introduce without drudgery or affectation.'

He presented the new sunspot evidence on Day Three of the *Dialogue*, the day devoted to the discussion of the Earth's annual motion. It came up right after a spirited demonstration of the planetary positions, in which Salviati showed how the errant wanderings of Mercury, Venus, Mars, Jupiter and Saturn could all be explained by the Earth's plying a yearly orbit, between the orbits of Venus and Mars, around the Sun.

'But another effect', Salviati adds without pausing for breath, 'no less wonderful than this, and containing a knot perhaps even more difficult to untie, forces the human intellect to admit this annual rotation and to grant it to our terrestrial globe. This is a new and unprecedented theory touching the Sun itself. For the Sun has shown itself unwilling to stand alone in evading the confirmation of so important a conclusion, and instead wants to be the greatest witness of all to this, beyond exception. So now hear this new and mighty marvel.'

And he proceeds to describe how the sunspots alter their apparent path, from the straight line that could be observed only two days out of every 365 (at the summer and winter solstices), to an arc that curves upwards for half the year and downwards for the other six months. Anyone who clung to the Ptolemaic system – who insisted the Sun spun around the Earth every day – now had to explain why the sunspots changed the angle of their path according to an annual cycle, and not a daily one. Some Aristotelian stalwarts might dredge up the old dismissal of the sunspots as vain illusions of the telescope lenses. More serious and scientific followers of Ptolemy, however, would have to spiral the Sun through the most complex gyrations imaginable, in order to save these newly recognised appearances in a geocentric and geostationary system.

Salviati's explication of the sunspot connection fills only ten pages of dialogue, including a couple of diagrams to help Sagredo and Simplicio grasp the concept. The three discussants thus find ample time on the third day to consider other imponderables – the size and shape of the firmament, for example – and to converse in awe about the confounding grandeur of space.

Responding to Simplicio's Aristotelian vision of the Earth as the centre of the world, the centre of the universe, the hub of the stellar sphere, Salviati proposes something vaster and vaguer: 'I might very reasonably dispute whether there is in Nature such a centre, seeing that neither you nor anyone else has so far proved whether the universe is finite and has a shape, or whether it is infinite and unbounded.'

This very modern idea of a universe without end had occurred to Copernicus but also helped kindle the fire that executed Giordano Bruno, as Galileo knew. He dropped the subject quickly yet returned again and again in the *Dialogue* to the evident enormity of the heavens.

Copernicus had pushed the stars away to unimaginable distances to explain their constancy in contrast to the planets. The reason the stars never seemed to rock this way or that as the Earth travelled all around the Sun over the course of the year, Copernicus explained, was that they lay too far away for any shift in position, or parallax, to be perceived. Galileo agreed, and furthermore predicted that improved observations through powerful future instruments would one day reveal this annual stellar parallax.*

Simplicio and his fellow Aristotelian philosophers really hated the big, unwieldy universe of Copernicus. They could not believe that God would have wasted so much space on something of no possible use to man.

'It seems to me that we take too much upon ourselves, Simplicio,' counters Salviati, 'when we will have it that merely taking care of us is the adequate work of Divine wisdom and power, and the limit beyond which it creates and disposes of nothing. I should not like to have us tie its hand so. . . . When I am told that an immense space interposed between the planetary orbits and the starry sphere would be useless and vain, being idle and devoid of stars, and that any immensity going beyond our comprehension would be superfluous for holding the fixed stars, I say that it is

---

* In 1838, German astronomer Friedrich Wilhelm Bessel finally discerned annual parallax for the star called 61 Cygni, thus demonstrating the Earth's orbital motion and gauging the greatness of the distance to even the nearest stars.

The operating theatre at the University of Padua

brash for our feebleness to attempt to judge the reason for God's actions, and to call everything in the universe vain and superfluous which does not serve us.'

The quick-witted Sagredo jumps in at this point, bristling that even the remote stars may serve a man in ways he cannot fathom. 'I believe that one of the greatest pieces of arrogance, or rather madness, that can be thought of is to say, "Since I do not know how Jupiter or Saturn is of service to me, they are superfluous, and even do not exist." Because, O deluded man, neither do I know how my arteries are of service to me, nor my cartilages, spleen or gall; I should not even know that I had gall, or a spleen, or kidneys, if they had not been shown to me in many dissected corpses.'

Sagredo's frequent anatomical analogies throughout the *Dialogue* recall that Andreas Vesalius published his revelations about human anatomy, *On the Fabric of the Human Body*, in 1543 – the same year as Copernicus's *On the Revolutions of the Heavenly Spheres*, and with just as much affront to Aristotle. Even while Galileo sat writing the *Dialogue* nearly a century later, Aristotelians still clung to the heart as the origin of the nerves, though Vesalius had followed their course up through the neck to the brain. Vesalius, who took his medical degree at Padua and lectured all over Italy, had also staged sensational popular demonstrations showing the male and female skeletons to contain the same number of ribs, thus defying the widespread belief, based on the Book of Genesis, that men came one rib short.

Although Galileo had left his own medical studies behind by the time he began teaching at Padua in 1592, he undoubtedly attended human dissections by torchlight in the university's tiered anatomical theatre, the world's first such facility, where the cadaver could be raised through a trapdoor in the floor from the canal below, and its body parts burned in a furnace after study. Several of Galileo's acquaintances on the Padua medical faculty, out of necessity, willed their own mortal remains to the university, to save the anatomists the bother of pillaging the hospitals or begging the bodies of criminals condemned to hang.

'Besides,' Sagredo sputters at last in frustration with those who would limit the majesty of the universe, 'what does it mean to say that the space between Saturn and the fixed stars, which these men call too vast and useless, is empty of world bodies? That we do not see them, perhaps? Then did the four satellites of Jupiter and the companions of Saturn come into the heavens when we began seeing them, and not before? Were there not innumerable other fixed stars before men began to see them? The nebulae were once only little white patches; have we with our telescopes made them become clusters of many bright and beautiful stars? Oh, the presumptuous, rash ignorance of mankind!'

Thus, to imagine an infinite universe was merely to grant almighty God His proper due.

[XVI]

The tempest
of our
many torments

ALL THROUGH THE AUTUMN of 1629, Galileo gave himself over to
the completion of the *Dialogue*, which he finished on Christmas
Eve. His health remained strong throughout this period, interrupt-
ing his three-month burst of creativity only once, in early November, when Suor Maria Celeste and Suor Luisa treated his brief
indisposition by sending him five ounces of their vinegary oxymel
concoction and some syrup of citron rind to ameliorate its bitter
taste.

Now that Sestilia had assumed the domestic care of Vincenzio
through her wedding vows, and also tended lovingly to many of
Galileo's personal needs, Suor Maria Celeste apparently found

another way to aid her father. One of her letters suggests that she busied herself recopying his draft manuscript of the *Dialogue*. Sections Galileo had composed at different times in different formats now needed to be handwritten page by page in perfect penmanship for publication, with corrections and additions pasted in as necessary. When Suor Maria Celeste referred to *ritagli* or 'clippings' already in her possession during November of 1629, reminding him of additional ones he had promised to send before she could set to work on the whole lot, she more than likely meant pieces of the *Dialogue* that reached her piecemeal. (The book's original manuscript has not survived, however, nor any description of it that notes whose handwriting it bore.)

On the short fourth – and final – day of the *Dialogue*, Galileo rehashed the 'Treatise on the Tides' he had given to Cardinal Orsini in 1616, in which he judged the flux and reflux of the sea to be the inevitable result of the Earth's two motions, one around its own axis, the other around the Sun: as the Earth turns, its daily spinning conspires with its annual revolution to jar the great oceans. Not only can the moving Earth account for the tides, Salviati indicates in his speeches, but also the tides, by their very existence, reveal the motion of the Earth.

After Simplicio denounces this idea as 'fictitious', the men continue to plumb the confounding complexity of the tides – how they vary in timing, in volume and in height from one part of the world to another. Salviati suggests that the effort of accounting for these anomalies in the Mediterranean alone led literally to the death of Aristotle: 'Some say it was because of these differences and the incomprehensibility of their causes to Aristotle that he, after observing them for a long time from some cliffs of Euboea, plunged into the sea in a fit of despair and wilfully destroyed himself.'

The denouement at that day's end, when time came for the three characters to close discussion and draw conclusions, demanded delicate diplomacy, for the text of the third and fourth days advanced compelling physical arguments in support of Copernicus, while the overall tenor of the book needed to preserve the spirit of hypothesis, as Galileo had promised Urban it would.

'In the conversations of these four days we have, then, strong evidences in favour of the Copernican system,' sums up the hospitable Sagredo, 'among which three have been shown to be very convincing – those taken from the stoppings and retrograde motions of the planets, and their approaches towards and recessions from the Earth; second, from the revolution of the Sun upon itself, and from what is to be observed in the sunspots; and third, from the ebbing and flowing of the ocean tides.'

But Salviati-cum-Galileo, although he has led the discussion in this direction, refuses to endorse Copernicus in the end. He concedes that 'this invention' – meaning the heliocentric design – 'may very easily turn out to be a most foolish hallucination and a majestic paradox'.

By such equivocation, Galileo offset his persuasive, often passionate defence of Copernicus. At the end of the *Dialogue*, he further sought to please Pope Urban by complying with His Beatitude's wish to see a reprise of the cicada philosophy from *The Assayer* – the idea that God and Nature possess boundless means for creating the effects observed by men. But Galileo placed these words in the mouth of Simplicio, where their profundity rang shallow: 'As to the discourses we have held, and especially this last one concerning the reasons for the ebbing and flowing of the ocean, I am really not entirely convinced,' balks the staunch Aristotelian,

> but from such feeble ideas of the matter as I have formed, I admit that your thoughts seem to me more ingenious than many others I have heard. I do not therefore consider them true and conclusive; indeed, keeping always before my mind's eye a most solid doctrine that I once heard from a most eminent and learned person, and before which one must fall silent, I know that if asked whether God in His infinite power and wisdom could have conferred upon the watery element its observed reciprocating motion using some other means than moving its containing vessels, both of you would reply that He could have, and that He would

have known how to do this in many ways which are unthinkable to our minds. From this I forthwith conclude that, this being so, it would be excessive boldness for anyone to limit and restrict the Divine power and wisdom to some particular fancy of his own.

And so they part company, hopeful of another fruitful convention on other fascinating themes at some future date.

As he neared completion of his *Dialogue*, Galileo anticipated submitting the final text for censorship. Not just sensitive subjects, such as the structure of the universe, but all books on any topics fell under this directive throughout Catholic Europe, following a papal bull issued in 1515 by the Medici pope Leo X. Writers seeking publication, this decree stated, must have their manuscripts scrutinised by a bishop of the Church or bishops' appointees, as well as by the local inquisitor. Printers who started their presses without the requisite permissions faced excommunication, fines and the burning of their books. For the special case of Germany, fount of the Reformation, Pope Leo perforce prepared another bull five years later, in 1520, prohibiting all works, past and future, from the pen of Martin Luther.

The Roman Inquisition, after its reorganisation in 1542, assumed supervision of printing projects in Italy, and in 1559 promulgated the first worldwide Index of Prohibited Books. In 1564, following the Council of Trent, harsher new restrictions stipulated that authors as well as printers could be excommunicated for publishing works judged heretical. Even the *readers* of such texts could be so punished. Booksellers, likewise, had to beware, keeping an exact listing of their stock, and standing ever ready for impromptu inspections called by bishops or inquisitors.

All of Galileo's previously published works had undergone the requisite scrutiny, for Italian printers obeyed the rules more strictly than most – especially in Rome, home of the Holy Office of the Inquisition. *The Starry Messenger*, printed in Venice, carried the approbation of the heads of the Venetian law-enforcement agency called the Council of Ten, as well as the resident 'Most

Reverend Father Inquisitor, the Overseers of Padua University, and the Secretary of the Venetian Senate', all of whom swore 'that in the book entitled *Sidereus Nuncius* by Galileo Galilei there is nothing contrary to the Holy Catholic Faith, principles, or good customs, and that it is worthy of being printed'.

When Prince Cesi was preparing to print the *Sunspot Letters* in Rome, he discussed with Cardinal Bellarmino the potential problem of impugning the Sun's incorruptibility,

Engraving of Galileo,
by Francesco Villamena

while Galileo double-checked this point with Cardinal Conti. Neither eminence thought the sunspots would disturb the censors, and in fact the book was licensed without incident.

The Assayer, too, had negotiated official channels smoothly. But Galileo suspected the substance of the *Dialogue* might give the censors serious cause for concern.

MOST BELOVED LORD FATHER

NOW THAT THE TEMPEST of our many torments has subsided somewhat, I want to make you fully aware of the events, Sire, without leaving anything out, for in so doing I hope to ease my mind, and at the same time to be excused by you, for dashing off my last two letters so randomly, instead of writing in the proper manner. For truly I was half beside myself, shaken by the terror aroused in me and in all of us by our novice mistress, who, overpowered by those moods or frenzies of hers, tried twice in recent days to kill herself. The first time she struck her head and face against the ground with such force that she became monstrously deformed; the second time she stabbed herself thirteen times, leaving two wounds in her throat, two in her stomach, and the others in her abdomen. I leave you to imagine, Sire, the horror that gripped us when we found her body all bloody and battered. But we were even more stupefied at how, as seriously injured as she was, she made the noise that drew us to enter her cell, asked for the confessor, and then in confession handed over to the priest the instrument she had used, so as to prevent any of us from seeing it (although, as far as we can conjecture, it was a pocket knife); thus it appears that she was crazy and cunning at the same time, and the only possible conclusion is that these are mysterious judgments of the Lord, Who still keeps her alive, when for every natural cause she should surely have died, as the wounds were all perilous ones, according to the surgeon; in the wake of these events we have guarded her continuously day and night. Now that the rest of us are recovered, by the grace of blessed God, and she is tied in her bed, albeit with the same deliriums, we continue to live in fear of some new outburst.

Beyond this travail of ours, I want to apprise you of another anxiety that has been weighing heavily on my heart. The very moment you were so kind as to send me the 20 *scudi* I had requested (I did not dare to speak freely of this in person, when you asked me recently if I had obtained the cell yet) I went with the money in my hand to find the nun who was selling it, expecting that she, being in extreme necessity, would willingly accept that money, but she simply could not resign herself to relinquishing the cell she loved so much, and since we did not reach an agreement between ourselves, nothing came of it, and I lost the chance to purchase that little room. Having assured you, Sire, that I could indeed obtain it, and then not succeeding, I became greatly troubled, not just on account of being deprived of my own space, but also because I suspected you would get upset, Sire, believing me to have said one thing and done another, though such deceit was never my intention; nor did I even want to have this money, which was causing me such grief. As it happened, the Mother Abbess was confronted at that point with certain contingencies, which I gladly helped her through, and now she, out of gratitude and kindness, has promised me the room of that nun who is sick, the one whose story I told you, Sire, whose room is large and beautiful, and while it is worth 120 *scudi* the Mother Abbess will give it to me for 80, thus doing me a particular favour, just as she has on other occasions always favoured me. And because she knows full well that I cannot pay a bill of 80 *scudi*, she offers to reduce the price by the 30 *scudi* that you gave the convent some time ago, Sire, so that with your consent, which I see no reason to doubt, as this seems to me an opportunity not to be missed, I will have all that I could ever want in the way of comfort and satisfaction, which I already know to be of great importance to you. Therefore I entreat your consideration, so that I can give some response to our Mother Abbess, who will be relinquishing her office in a few days, and is currently settling her accounts.

I also want to know how you feel, Sire, now that the air is slightly more serene, and, not having anything better to send you, I offer a little poor man's candied quince, by which I mean that I prepared it with honey instead of sugar, so if it is not right for you, perhaps it will satisfy the others; I would not know what to give my Sister-in-law now, in her condition [pregnant Sestilia was near term]. Surely if she had a taste for anything made by nuns, Sire, you would tell us, because we want so much to please her. Nor have I forgotten my obligation to La Porzia [Galileo's housekeeper], but circumstances have prevented me from making anything as yet. Meanwhile if you have gathered the additional clippings you promised me, Sire, I will be very happy to receive them, as I am holding off work on those I already have until the others arrive.

I must add that, as I write, the sick nun I mentioned earlier has taken such a turn that we think she is on the verge of death; in which event I will be obliged to give the remainder of the money to Madonna right away, so that she can make the necessary purchases for the funeral.

In my hands I hold the agate rosary you gave me, Sire, which is excessive and vain for me, while it seems perhaps right for my Sister-in-law. Let me therefore return it to you, so you can learn if she would like to have it, and in exchange send me a few *scudi* for my present need, so that, if it please God, I believe I really will have the full sum; and in consequence I will no longer be forced to burden you, Sire, for that is what concerns me most. But in fact I do not have, nor do I want to have, others to whom I can turn, except for you and my most faithful Suor Luisa, who wearies herself doing everything she can for me; but in the end we depend upon each other because alone we lack the strength that circumstances so often demand of us. Blessed be the Lord Who never fails to help us; by Whose love I pray you, Sire, to forgive me if I vex you too much, hoping that God Himself will reward you for all the good things you have done for us and continue to do, for which I thank you with all my heart, and I entreat you to excuse me if you find any errors here, because I do not have time to reread this long litany.

FROM SAN MATTEO, THE 22ND DAY OF NOVEMBER 1629. Your most affectionate daughter,

S. Maria Celeste

Hail Mary, full of grace, the Lord is with thee. Blessed art thou among women, and blessed is the fruit of thy womb, Jesus.

Holy Mary, Mother of God, pray for us sinners, now and at the hour of our death.

The rapt repetition of the Franciscan Crown Rosary – the seventy-three Hail Marys, the six Our Fathers, the contemplation of the Joyful Mysteries – required no gilding of agate for a Poor Clare. Wooden beads would work just as well, or even chains of dried, hardened rosebuds.

Our Father, Who art in Heaven, hallowed be Thy name. Thy kingdom come, Thy will be done, on Earth as it is in Heaven.

Give us this day our daily bread and forgive us our trespasses as we forgive those who trespass against us. Lead us not into temptation but deliver us from evil.

Galileo gave Suor Maria Celeste the money for her need right away. And early in December Sestilia gave Vincenzio a son – the third and last Galileo Galilei.

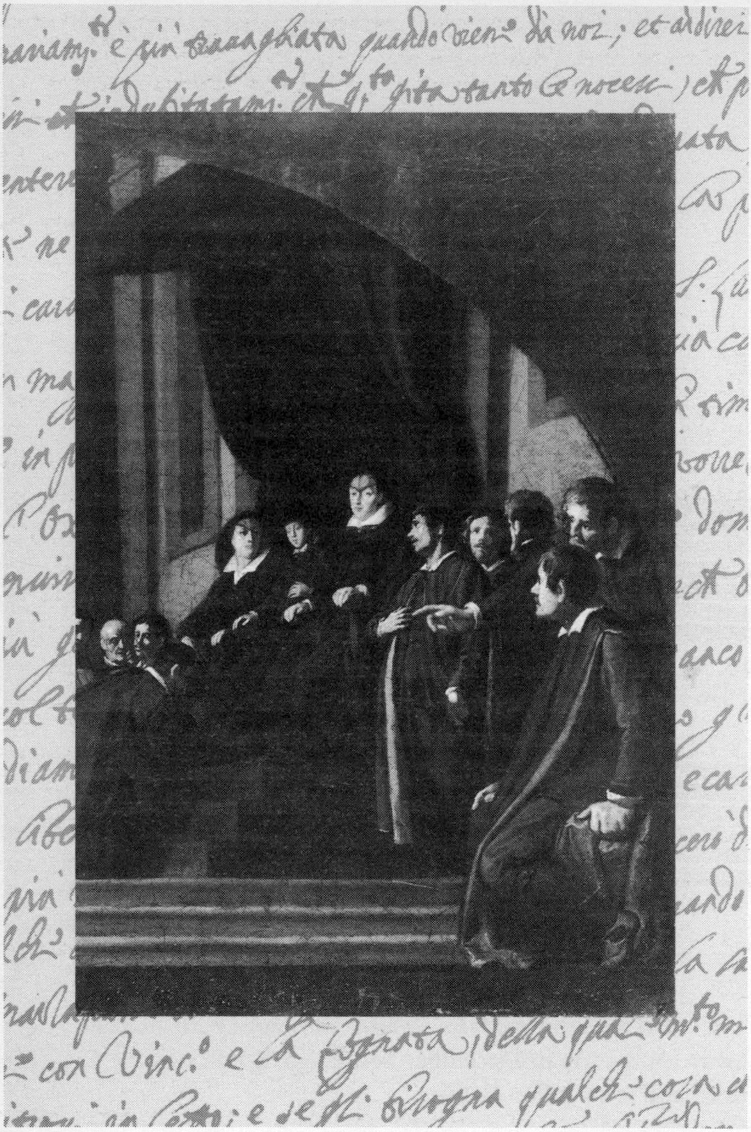

PART THREE

In
Rome

[XVII]

While seeking to immortalise your fame

BY THE TIME GALILEO finished writing his book about the world systems, just as December 1629 drew to a close, he had established a new closeness with his daughter. Ever a source of love and financial aid to Suor Maria Celeste, as well as the grateful recipient of her labours, he now began to do favours for her that required the skilled work of his own hands. And she, emboldened either by her recent assistance on the manuscript for the *Dialogue* or by the maturity of her nearly thirty years, engaged her father with increased confidence. Before too long, the strength of their mutual affection and deepening interdependence would submit to tests neither one of them could yet imagine.

Convent clock of the type used at San Matteo

Having moved into her private room, Suor Maria Celeste found it spacious enough to accommodate a small party of sisters at afternoon needlework. The single, small, high window, however, admitted only a dim light, and so she asked Galileo if she might send him the window frames to be refitted with newly waxed linen. 'I do not doubt your loving willingness in this matter,' she said of her request, 'but the fact that the work is rather more suited to a carpenter than a philosopher gives me pause.'

She also prevailed on him several times to repair the convent's temperamental clock. He fixed it once when its chime failed to wake the sacristan (who in turn failed to summon the other sisters from sleep to the Midnight Office) and again whenever it developed a different quirk. 'Vincenzio worked on our clock for a few days, but since then it sounds worse than ever,' she told Galileo on 21 January 1630. 'For my part, I would judge the defect to be in the cord, which, owing to its being old, no longer glides. Still, as I am unable to fix it, I turn it over to you, so that you can diagnose its deficiency, and repair it. Perhaps the real defect was with me, in not knowing the right action to take, which is the reason I have left the counterweights attached this time, suspecting that perhaps they are not in their proper place; in any event I beseech you to send it back as quickly as you possibly can, because otherwise these nuns will not let me live.'

Galileo's brother, Michelangelo, had purchased the convent's portable clock, which stood about two feet tall, in Germany. Like all mechanical timekeepers of its day, this one offered no precision improvement over the sundial, though it did mark time through the dark or the rain and strike the hour aloud.

Italians numbered the hours of the seventeenth century from one to twenty-four, beginning at sunset, so that if Suor Maria Celeste told her father she was 'writing at the seventh hour', she meant she worked far into the night. And when she reported the death of an ailing nun 'at the fourteenth hour' of a November Sunday, the time indicated was about half past six in the morning.

'The clock that travelled back and forth between us so many times now runs beautifully,' Suor Maria Celeste said in a thank-you note on 19 February, 'its flaw having been my fault, as I adjusted it improperly; I sent it to you in a covered basket with a towel, and have not seen either of these since; if you find them by chance about your house, Sire, please do return them.'*

As he performed these services, Galileo initiated the rigmarole of licensing and printing his recently completed book. Since Prince Cesi of the Lyncean Academy intended to publish the *Dialogue* in Rome, the work would have to undergo censorship there in the Holy City, despite the fact that its author lived in Florence. Galileo, now almost sixty-six, planned personally to deliver the manuscript to the relevant authorities at the Vatican. But Rome was a distant country, and an old man risked his life adding wintry weather to the perils inherent in a two-hundred-mile journey.

In February, while Galileo waited for spring, Pope Urban VIII unexpectedly issued a formal salute to his 'honest life and morals and other praiseworthy merits of uprightness and virtue'. With these words Urban gave Galileo a prebend in Pisa – similar but

---

* Galileo's efforts here increased his familiarity with clockwork and helped lead him to his invention, some ten years later in 1641, of a prototype pendulum clock. Vincenzio helped his father by drawing a blueprint and building a model, but the work was never completed by either of them. When Christiaan Huygens later patented a pendulum clock in Amsterdam in 1656, Galileo's followers accused Huygens, albeit unjustly, of plagiarism.

unrelated to the previously granted canonry at Brescia, which had bounced from Vincenzio to Vincenzio and then out of the Galilei family bounds. Rather than accept the Pisan prebend right away, however, Galileo tried instead to reclaim the Brescian one, now that its incumbent had died, for his infant grandson.

'I do not think it would be possible to confer this pension on a baby without a dispensation which will be very difficult to obtain,' Galileo's old friend Benedetto Castelli counselled him on the matter. (After nearly a year's manoeuvring, Galileo himself emerged as canon of Brescia *and* canon of Pisa – with a combined annual annuity of one hundred *scudi* from Church revenues. While his clerical posts did not require him to wear a habit or change his lifestyle, he did have his head shaved in an ecclesiastical tonsure by the bishop of Florence.)

Mid-March turned Galileo's thoughts towards Rome and his impending departure to procure the printing licence for the *Dialogue*. His typical devotion to scholarly pursuits at this point raised the usual concerns in Suor Maria Celeste. 'And I would not want you, while seeking to immortalise your fame,' she fretted in an April letter, 'to cut short your life; a life held in such reverence and treasured so preciously by us your children, and by me in particular. Because, just as I have precedence over the others in years, Sire, so too do I dare to claim that I precede and surpass them in my love for you.'

Galileo gathered the whole family, including Vincenzio and the again-pregnant Sestilia, at the convent's parlour grille on the morning of 15 April to say goodbye. By 3 May, he had arrived at the Tuscan embassy in Rome, where he lodged for the next two months as the guest of Ambassador Francesco Niccolini and his charming wife, Caterina. Not only did Galileo enjoy their hospitality, but he also prized their connections at the Vatican: Her Ladyship the Ambassadress was a close cousin to the Dominican father Niccolò Riccardi, who controlled the Roman licensing of books. Technically Pope Urban, as bishop of Rome, owned this and all other ecclesiastic powers pertaining to the stewardship of the city. But since the pope's diocese encompassed the world, he delegated most local affairs to his cardinal vicar, and the censorship of books to his master of the Sacred Palace. Father Riccardi bore this important title, although his cousins at the embassy affectionately called him 'Father Monster' – a nickname he had earned from King Philip III of Spain, in recognition of his imposing physical proportions and mental prowess.

Galileo could not have handpicked a more favourable judge than Father Riccardi. The man came from a fine Florentine family closely allied with the Medici, and he had greeted Galileo's previous work, *The Assayer*, with a gush of admiration. 'Besides having found here nothing offensive to morality,' Father Riccardi stated in *The Assayer*'s imprimatur in 1623, 'nor anything which departs from the supernatural truth of our faith, I have remarked in it so many fine considerations pertaining to natural philosophy that I believe our age is to be glorified by future ages not only as the heir of works of past philosophers but as the discoverer of many secrets of nature which they were unable to reveal, thanks to the deep and sound reflections of this author in whose time I count myself fortunate to be born – when the gold of truth is no longer weighed in bulk and with the steelyard, but is assayed with so delicate a balance.'

Ambassador Niccolini nevertheless took a grim view of the road ahead. He sent word to the grand duke in Florence that he expected trouble because of Galileo's numerous, vociferous enemies, particularly among the Jesuits, who were already criticising both the premise of the new book and its author. Reliable rumours ran through Rome, alleging how the *Dialogue* explained the tides by the motion of the Earth. Theologians naturally frowned on such a notion.

Father Riccardi read the book himself. He also deputised a fellow Dominican with expertise in mathematics to scrutinise the text and report back to him. Meanwhile, Galileo facilitated a friendship by correspondence between his elder daughter and his kind hostess, Caterina Riccardi Niccolini, for he saw Her Ladyship the Ambassadress as a potential patroness for the sisters of San Matteo.

'The Mother Abbess sends her regards to you, Sire,' Suor Maria Celeste wrote at the end of May,

> and reminds you of what she told you in person: that is, if chance should offer you the opportunity there in Rome to obtain some charitable help for our Monastery, to please extend yourself in this effort for the love of God and for our relief; although I must add that truly it seems an extraordinary thing to ask of people living so far away, who, when doing a good deed for someone, would prefer to favour their own neighbours and compatriots. Nonetheless I know that you know, Sire, by biding your time, how to pick the perfect moment for implementing your intentions successfully; and therefore I eagerly encourage you in this endeavour, because indeed we really are in dire need, and if it were not for the help we have received from several donations of alms, we would be at risk of starving to death.

During the eight weeks Galileo spent in Rome, Pope Urban granted him only one gracious audience. His Holiness had no time for more. The past several years had enmeshed him in difficult papal affairs, including political machinations pertaining to the Thirty Years' War. What had begun as a battle between German Catholics and German Protestants had pulled into its vortex kings and princes of many other countries, including France, Spain, Portugal, Denmark, Sweden, Poland, Transylvania and Turkey – all contributing to the carnage on German soil. By 1630, only a few of the multiple causes fuelling the ongoing fighting still pertained to issues of religious faith; one of these was the struggle between the Catholic royal families of France and Spain for control of the Catholic throne of the Holy Roman emperor in Germany. The pope, as the leader of all Catholics everywhere, might have tried harder to unite the Spanish Habsburgs and the French Bourbons. But instead, Urban, who had served early in his Church career as papal legate in France, where he held the newborn Louis XIII at the baptismal font, allied himself with King Louis and Cardinal

Richelieu. Of late, Urban had grown so fearful of Spanish spies in the Vatican, he complained, that he dared not speak above a whisper. The pontiff's war worries so disrupted his sleep that he ordered all the birds in his gardens killed, lest they offend him with nocturnal calls.

In addition to the Thirty Years' War, Urban entered the War of the Mantuan Succession, which broke out in northern Italy in 1627. In the strategically located Duchy of Mantua, in Lombardy, an aged duke and his brother both met their deaths without having produced the obligatory male heir.* In 1625, the dying duke had installed a French relative in his Italian office, hoping to keep the property in the family, but a bloody dispute erupted over Mantua that reflected the general tension between France and Spain for an Italian foothold. Urban contributed military support to Mantua and made new enemies among the citizens of Rome by taxing them to help pay the high cost of equipping seven thousand infantrymen and eight hundred cavalry.

Meanwhile, Austrian Habsburg troops fighting in Mantua inadvertently released a biological weapon along with their musket fire, by carrying the bubonic plague across the Alps into Italy in 1629. Urban, as part of his programme to protect the populace from this threat, now travelled clear across Rome every Sunday to say mass at the Church of Santa Maria Maggiore, where a treasured image of the Madonna had miraculously barred the plague from the city in the sixth century.

Adding to Urban's woes, an astrological forecast in the spring of 1630 prophesied his own early demise. The superstitious pontiff retaliated first with imprisonment for the astrologer and later with a ferocious edict prohibiting predictions of a pope's death, or even the deaths of papal family members up to and including the third degree of consanguinity. Urban's consistent concern for the male members of the Barberini line manifested itself in a perpetual stream of promotions and pensions. The pope had all but dragged

---

* Galileo had tried to attach himself to the court of Vincenzo Gonzaga at Mantua, in 1603, but the salary the duke offered him fell short of his professor's pay at Padua, so he stayed put until he secured a better position with the Medici.

his brother Antonio, a Capuchin monk, away from the simple devotional life at the monastery to become a cardinal in the papal court in 1624. He later made another cardinal of his nephew Antonio in 1628, though the youth (the baby brother of the cardinal nephew Francesco) was only nineteen. Urban designated the middle child among his three nephews, Taddeo, to father the next generation of Barberinis and married him into a family of landed, titled Roman gentry.

Urban had continued all the while to indulge his own intellect by writing and publishing epic poetry. He had also reformed the breviary – the religious orders' handbook of prayers, offices, hymns, saints and instructional passages from the Bible – by adding his own original hymnal compositions in honour of the saints he had canonised. And then, wishing to bring the existing, mostly ancient hymns in the old breviary up to his high standards, Urban had commissioned a cadre of literary Jesuits to edit them for grammar and metre.

He had further sought to ensure the immortality of his name through the construction of an ornate bronze canopy above the tomb of Saint Peter inside the Vatican basilica. In 1624, soon after his election, Urban charged the sculptor Gian Lorenzo Bernini to begin building this *baldacchino*, or arch, over the altar where only the pope himself was permitted to preside. By the time of Galileo's 1630 visit to Rome, the *baldacchino*'s mammoth form was rising inside the church from four marble cornerstones, each emblazoned on two sides with the Barberini three-bee coat of arms. Four support pillars spiralled ninety-five feet up towards the canopy, to be crowned by a cross of gold and a coterie of carved angels. The bronze pillars contained beams lifted from the Pantheon, an enormous ancient structure that had survived the sack of Rome by barbarians in the fifth century. When Urban ordered the venerable Pantheon looted for his armaments' and monument's sake, outraged Romans plaintively played on his name: 'What the barbarians did not do', they cried, 'the Barberini did.'

Everywhere Galileo looked in Rome, he saw evidence of Urban's aggrandisement. Although the Tuscan philosopher remained in the pope's favour – as the recent prebend and the private audience attested – he could no longer expect the same degree of personal contact that had been his privilege in the past. Galileo saw no reason to feel slighted by this change, however, especially when invited to dine as the special guest of the cardinal nephew, and when told how Father Riccardi was discussing the details of his *Dialogue* in closed session with His Holiness.

Baldacchino designed by Bernini for Pope Urban VIII

On 16 June Galileo found out that the *Dialogue* had passed the reviewer's inspection. Within days Father Riccardi signed the manuscript, giving Galileo a provisional licence to publish in exchange for the promise of a few corrections. The title, for example, displeased the pope with its allusion to the physical phenomenon of the tides. Galileo had to choose a more mathematical or hypothetical title. Also the preface and ending must support the pope's philosophy of science, which attributed all complexity in Nature to the mysterious omnipotence of God. In any event, preliminary negotiations with printers regarding the *Dialogue* could now begin.

Satisfied, Galileo rushed to leave the city by the end of June, before the plague or malaria wafted into Rome on the summer heat. He promised Father Riccardi he would write the necessary corrections back home at Bellosguardo and then return with the revised manuscript in the autumn.

'Had the season been different I would have stayed to have it printed', he explained later that summer to a correspondent in Genoa, 'or else left it in the hands of Prince Cesi, who would take care of it as he has done other works of mine; but he was ill, and what is worse I now understand he is near death.'

Indeed, the good Prince Cesi passed away on 1 August, at forty-five, killed by gangrene of the bladder. His loss all but dissolved the Lyncean Academy. With Cesi gone, Galileo realised he would have to supervise publication of the *Dialogue* himself. And in that case he hoped to avoid trekking back and forth to Rome any longer, but would try instead to procure a licence from local authorities in Tuscany.

Before he could put this plan into effect, however, the deadly plague that had spread through Milan and Turin over the past year moved south to invade the city of Florence. Soon, stricken families on both sides of the Arno mourned new victims of the pestilence at every hour. Up on the hill at Bellosguardo, one of Galileo's craftsmen, a glassblower, died with his skin discoloured by blackish splotches and a putrid bubo festering in his groin.

[XVIII]

Since the Lord chastises us with these whips

MOST BELOVED LORD FATHER

I AM HEARTSICK AND WORRIED, Sire, imagining how disturbed you must be over the sudden death of your poor unfortunate worker. I assume that you will use every possible precaution to protect yourself from the danger, and I fervently urge you to make great effort in this endeavour; I further believe that you possess remedies and preventatives proportionate to the present threat, wherefore I promise not to dwell on the subject. But still with all due respect and filial confidence I will exhort you to procure the best remedy of all, which is the grace of blessed God, by means of a thorough contrition and penitence. This, without doubt, is the most efficacious medicine, not only for the soul, but for the body as well: since, given that living happily is so crucial to the avoidance of contagious illness, what greater happiness could one secure in this life than the joy that comes of a clear and calm conscience?

It is certain that when we possess this treasure we will fear neither danger nor death; and since the Lord justly chastises us with these whips, we try, with His aid, to stand ready to receive the blow from that mighty hand, which, having magnanimously granted us the present life, retains the power to deprive us of it at any moment and in any manner.

Please accept these few words proffered with an overflowing heart, Sire, and also be aware of the situation in which, by the Lord's mercy, I find myself, for I am yearning to enter the other life, as every day I see more plainly the vanity and misery of this one: in death I would stop offending blessed God, and I would hope to be able to pray ever more effectively, Sire, for you. I do not know but that this desire of mine may be too selfish. I pray the Lord, who sees everything, to provide through His compassion what I fail to ask in my ignorance, and to grant you, Sire, true consolation.

All of us here are in good physical health, save for Suor Violante, who is little by little wasting away: although indeed we are burdened by penury and poverty, which take their toll on us, still we are not made to suffer bodily harm, with the help of the Lord.

I am eager to know if you have had any response from Rome, regarding the alms you requested for us.

Signor Corso [Suor Giulia's brother] sent a weight of silk totalling 15 pounds, and Suor Arcangela and I have had our share of it.

I am writing at the seventh hour: I shall insist that you excuse me if I make mistakes, Sire, because the day does not contain one hour of time that is mine, since in addition to my other duties I have now been assigned to teach Gregorian chant to four young girls, and by Madonna's orders I am responsible for the day-to-day conducting of the choir: which last creates considerable labour for me, with my poor grasp of the Latin language. It is certainly true that these exercises are very much to my liking, if only I did not also have to work; yet from all this I do derive one very good thing, which is that I never ever sit idle for even one quarter of an hour. Except that I require sufficient sleep to clear my head. If you would teach me the secret you yourself employ, Sire, for getting by on so little sleep, I would be most grateful, because in the end the seven hours that I waste sleeping seem far too many to me.

I shall say no more so as not to bore you, adding only that I give you my loving greetings together with our usual friends.

FROM SAN MATTEO, THE 18TH DAY OF OCTOBER 1630.
Your most affectionate daughter,

S. M. Celeste

The little basket, which I sent you recently with several pastries, is not mine, and therefore I wish you to return it to me.

Suor Maria Celeste's prescription for prayer and penitence in the face of the plague meshed perfectly with prevailing wisdom.

Prayer surpassed or at least augmented the many available treatments, which included blood-letting, crystals of arsenic applied to the wrists and temples, small sacks of precious stones laid over the heart, and unguents made by cooking animal excrement together with mustard, crushed glass, turpentine, poison ivy and an onion. The value of confession and penitence reached new heights during plague epidemics, for the disease, once it struck, left its victims no time to make amends.

The first symptom typically erupted as a swelling of the lymph nodes under the arms or between the thighs. These large, painful, pus-filled lumps, called buboes, gave the pestilence the name 'bubonic plague'. Ranging in size from almonds to oranges, they were the focus of treatment by doctors, some of whom advocated burning the buboes with incandescent gold or iron, then covering the wound with cabbage leaves; others preferred to cut the buboes open with a razor, suck out the blood, and deposit three leeches on the site, topped by a quartered pigeon or a plucked cockerel. Left alone, the buboes enlarged each day until they often burst on their own, provoking agony sufficient to rouse even the nearly dead to frenzy.

Since only a fraction of those who contracted the plague could hope to recover, the appearance of the bubo pronounced doom. Fevers rose high within hours of onset, accompanied by vomiting, diffuse pain that felt like burning or prickling, and delirium. Soon the skin displayed weals and dark markings caused by subcutaneous haemorrhage. Death followed within the week, unless the disease also invaded the lungs, where it caused the coughing of a bloody, frothy sputum and killed the victim in two or three days flat – but not before it facilitated numerous new infections via droplets borne on the wind to be inhaled by the unsuspecting.

The plague leaped like fire from the sick to the whole. Even the discarded clothes or other belongings of the afflicted could communicate the disease. When one family member fell sick, the rest typically followed and soon all were buried in the fields, as the laws of Florence barred plague corpses from the churchyards.

The Italian people recognised the plague as an ancient enemy, periodically banished but never vanquished. In its worst visitation, from 1346 until 1349, it claimed 25 million lives, or roughly one-third of the population of Europe, North Africa and the Middle East.

'Oh, happy posterity', exclaimed poet Francesco Petrarca when that Black Death robbed him of his beloved Laura, 'who will not experience such abysmal woe and will look upon our testimony as a fable.'

Following the pandemic of the fourteenth century, the plague returned to one region or another at its own whim every few years, as though to remind the wayward of the tortures of Hell.

By the early seventeenth century, Europeans had gained enough experience with the pestilence to recognise the accumulation of dead rats in streets and houses as the harbinger of disease. The causal connection, however, remained elusive. People continued to blame the plague on miasmas of swampy air, the full Moon, conjunctions of the planets, famine, fate, beggars, prostitutes or Jews. Two hundred years before the germ theory of disease, no one realised that the plague was caused by microbes living in and on the ubiquitous black rats.* When a sick rat died, its hungry fleas jumped the few inches to another animal, or to a nearby human. Having ingested infected blood, the fleas delivered the disease by inoculation with their next bite. The poisonous plague bacterium multiplied rapidly in a new host's bloodstream until infection pervaded the body, attacking vital organs to cause kidney failure, heart failure, haemorrhaging blood vessels and death by septic shock. (Galileo's early description of the flea he observed through his microscope as 'quite horrible' referred only to its ugliness, and not its true menace, for he had no inkling of the creature's role in plague transmission.)

While live rats travelled with impunity on foot and aboard ship, exporting the plague hither and yon, people were confined to their houses by fear and municipal prohibitions. The Venetian doge

---

* The microbe was finally identified in 1894 by French bacteriologist Alexandre Yersin of the Pasteur Institute.

imposed the first official quarantine in 1348, after the city death rate from plague had risen to six hundred per day. The doge's council, seeking to isolate returning voyagers from the Orient, had decided on the duration of forty days – *quaranta giorni* in Italian, from which the word *quarantine* is derived – by selecting the same period of time that Christ had sequestered himself in the wilderness.

When the pestilence returned in 1630, it skirted Venice altogether, dealing its most brutal punishment to the city of Milan and its environs in Lombardy. Nevertheless, the presence of the threat, coupled with the spectre of past plagues, mobilised preventive measures all over the peninsula. In Rome, for example, Pope Urban VIII led holy processions and granted indulgences to individuals who followed a formula of church visits, fasting, prayer, the giving of alms and the taking of communion. He installed his capable nephew, Francesco Cardinal Barberini, as head of the Congregation of Health, which issued policies and edicts to be enforced by civic authorities. Specially appointed guards stationed at the twelve gates through the city walls detained all travellers who arrived from plague regions.

The plague of 1630 offered the grand duke of Tuscany, Ferdinando II, his first chance to assert his new leadership. The twenty-year-old Medici scion had only just attained his majority in 1628, at which point he reclaimed the government from his mother, his grandmother and the Council of Regency. He responded with energy to the present emergency. In addition to all his official acts aimed at contagion control, Ferdinando went out among the people every day to show his compassion, walking the streets through the poorest neighbourhoods to dispense comfort and encouragement.

The portly Ferdinando looked nothing like a hero. He assumed a suit of armour only to pose for the court portraitist and never wore one into battle. Even on his hunting excursions, he left the bloody work to the falcons. But Ferdinando behaved fearlessly in the face of the plague. In contrast to the many well-to-do Florentines who fled the city in dread, he stood fast throughout the epidemic. He made his four younger brothers stand by him, lending their help, though the Medici family of course possessed the means to flee and many country estates to flee to – some complete with moat and drawbridge. As though rewarded for valour, Ferdinando and his seven siblings all survived the plague in Florence.

Grand Duke Ferdinando II

Ferdinando ceded broad discretionary powers to his public health officers – a group of noblemen who answered directly and only to him. Their ordinances, aimed with good intentions at halting the spread of infection, affected every aspect of daily life. Citizens who resented the policing of their private affairs found ways around the law, and the commissioners found themselves sometimes taunted in the streets, pelted by stones or formally denounced. For example, the Tuscan clergy, outraged by restrictions on gathering the multitudes together for public sermons and processions, appealed to Pope Urban VIII for redress. The pontiff responded by censuring the whole board of health – which included Galileo's friend and former student Mario Guiducci – so that the board members were forced to perform salutary penance for their good deeds.

Hired monitors working for the Florentine Magistracy of Public Health patrolled the city, sending the stricken to the plague hospitals, burning their belongings, and disinfecting and boarding up their homes – with the other family members locked inside. Behind these sealed, marked doors, the relatives were expected to wait twenty-two days for release while subsisting on allotments of bread and money distributed by the magistracy. The confined raised these supplies from the street in baskets on ropes and lowered their corpses from the windows via the same system.

Most often the man of the house, at the first hint of 'the bad

Plague doctor in protective costume; the beak was filled with flowers to ward off plague vapours

disease' among the children, would move a mattress and some cooking utensils into his shop, hoping to avoid enclosure and carry on business as usual through the epidemic. When an artisan or apprentice fell sick, his fellow workers might care for him themselves, often breaking the law by failing to report his case and bribing a barber-surgeon or a herbalist for secret treatment. Though all parties faced *strappado* torture (being hung by the wrists with hands tied behind their backs) if discovered in such behaviour, still they took the risk. Well-meaning weavers, gold beaters and butchers managed to foil the magistracy – not just by virtue of their innate cleverness but because the inadequate army of health deputies could not catch every infraction. Also the eleven hundred health workers, endowed with no special immunity, succumbed to the plague as readily as anyone else, constantly necessitating the appointment of new replacements from the ever-shrinking pool of available able-bodied men.

In defiance of the public health edicts, rich and poor alike often tried to hide their sick in the bosom of the family, rather than relinquish them to the isolation of the hospital. If a blackened child died at home, at least the intimacy of the tragedy offered its own solace – and held out hope of a finagled death certificate that could convince grave diggers and sacristan to bury the body in the churchyard.

A grim new sound insinuated itself into the daily clamour of the city districts – a sound that could still be heard at night when curfews kept the populace indoors. It was the ringing of the bells worn about the ankles of body removers and other sanitary personnel.

Meanwhile, the dreaded plague-houses at San Miniato and San Francesco took in some six thousand Florentines who bore buboes or ran suspicious fevers during the autumn of 1630. Most of the patients admitted to these institutions died in them, despite the efforts of resident doctors and clergy, and wound up thrown naked without ceremony into mass graves bordering the city. The magistracy spread lime and erected fences around the graveyards to keep dogs from devouring the cadavers, but the image of some beast bringing home the bones of loved ones now deceased became the communal nightmare.

Much to Galileo's distress, Vincenzio and the pregnant Sestilia joined the many who sought safety in rural settings – as though the wrath of God would not pursue them beyond the city walls. The couple removed to a villa the Bocchineri family owned at Montemurlo, in the hinterlands between Prato and Pistoia. Abandoning their house on the Costa San Giorgio, they also left behind their little boy, not yet one year old, in the care of a neighbourhood wet nurse and his grandfather Galileo.

[XIX]

The hope of having you always near

THE SISTERS OF SAN Matteo survived the autumn of the plague year without a single occurrence of the bad disease within the convent's enclosure. In addition to their prayers, they benefited from the strict precautions spelled out for religious residences by the Magistracy of Public Health. These stipulated, for example, that only the nuns' closest relatives could venture near enough to speak to them, and in any case the grille had to be covered with a sheet of parchment or paper during all such encounters.

'I feel there is no question of your coming here to see us', Suor Maria Celeste wrote to Galileo under the circumstances, 'and I do not ask you to visit, because none of us could derive much pleasure from it, unable as we are to converse freely just now.'

For further security, the convent steward and his wife were

forbidden to enter any houses in town or even attend mass at another church without written permission from the mother abbess. Incoming letters had to be grasped by forceps and then tanned with acid or bathed in vinegar or held near a flame to purge them of possible contaminants. Any food procured from outside the convent's own kitchen garden had to be purchased early in the morning, before it had a chance to be handled by too many shoppers in the market-place. When the steward took the convent's grain to the mill, he was enjoined to guard it throughout the grinding, making sure that no other grain got mixed in with it and no one else so much as touched it, then collect the finished flour in the same sack he came with and carry it straight back to be baked into bread immediately.

None of these measures – nor any of the many other safeguards regarding every detail from the distribution of chaplains' vestments to the laundering of nuns' blankets – would have barred rats from the premises, to be sure, but the sisters' extreme scarcity of food, which reached a critical nadir during the plague, perhaps made the convent walls not worth the bother of breaching.

The only death recorded at San Matteo all through 1630 was the late November passing of the long-suffering Suor Violante, whose onset of fever and dysentery pre-dated the plague epidemic by many months. In the outside world, however, the evil pestilence claimed Suor Maria Teodora's brother Matteo Ninci, a lad loved by all who knew him, and presently Suor Maria Celeste grieved to learn that the plague in Germany had killed her uncle Michelangelo.

Cold, hungry and surrounded by danger, the nuns appealed for aid from a variety of sources via a letter-writing campaign. Suor Maria Celeste wrote the letters, and Madonna – Madre Caterina Angela Anselmi, the current abbess of San Matteo in Arcetri – signed them. Their supplications began to bear fruit on All Saints' Day (1 November), when the dowager grand duchess, Madama Cristina, saw an abundance of bread delivered to the convent, followed soon thereafter by eighteen bushels of grain, and later by cash contributions from Their Serene Highnesses in excess of two hundred *scudi*.

The archbishop of Florence also pledged his aid. He demanded a list of names of the sisters' family members, with an eye towards asking them for financial support through the coming wintertime of want. Suor Maria Celeste panicked that such a move would tax Galileo at an unfair rate. Her father already gave generously, and now he would be pressed to pay not only his own share of the archbishop's new demands but also Vincenzio's, and perhaps other relatives' as well. Rather than let this happen, she devised a way to restructure the finances of San Matteo, so that sums of money long held privately by certain sisters' families – and all the interest that had accrued over the years – might now be called in to meet the immediate needs of the convent. Since she could not dictate strategy to the archbishop, however, she proposed this idea to Galileo, for him to present as his own.

'Here I cannot say more,' she apologised, 'except to call this affair to the attention of the Lord God, and entrust the rest, Sire, to your prudence.' This course eventually resulted in the desired outcome.

To help him in turn, Suor Maria Celeste addressed Galileo's personal and professional disappointments with prayers and exhortations. 'I beseech you not to grasp the knife of these current troubles and misfortunes by its sharp edge, lest you let it injure you that way,' she wrote in early November, 'but rather, seizing it by the blunt side, use it to excise all the imperfections you may recognise in yourself; so that you rise above the obstacles, and in this fashion, just as you penetrated the heavens with the vision of a Lyncean, so will you, by piercing also through baser realms, arrive at an awareness of the vanity and fallacy of all earthly things: seeing and touching with your own hands the truth that neither the love of your children, nor pleasures, honours or riches can confer true contentment, being in themselves ephemeral; but only in blessed God, as in our final destination, can we find real peace. Oh what joy will then be ours, when, rending this fragile veil that impedes us, we revel in the glory of God face to face!'

Galileo, though frustrated by his continuing inability to publish his *Dialogue*, began to write a new book based on the studies of motion he had begun long ago in Padua. He also kept up his scientific correspondence and consulted regularly on projects for the grand duke. In December, for example, Ferdinando asked Galileo to decide which of two rival engineers' schemes would best keep the Bisenzio River from overflowing its banks, and later bade him advise an architect on plans for repairing the façade of Florence's signature cathedral, Santa Maria del Fiore.

'I will go on taking good care of myself, as you urge me to do,' Suor Maria Celeste promised her hardworking father. 'And how I do desire you to heed some of the same advice you offer me, by not immersing yourself so deeply in your studies that you jeopardise your health too markedly; for if your poor body is to serve as an instrument capable of sustaining your zest for understanding and investigating novelties, it is well that you grant it some needed rest, lest it become so depleted as to render even your powerful intellect unable to savour that nourishment it devours with such relish.'

Suor Maria Celeste further nursed Galileo by plying him with every new plague preventive she could fabricate in her apothecary shop or procure by other means. Although she did not explain how she managed to obtain it, she sent him a bottle of healing water from the venerated Abbess Ursula of Pistoia – a liqueur that Ursula's own followers placed on a par with holy relics and dared not distribute to outsiders. 'Therefore I urge you, Sire, to put your faith in this remedy, because if you believe as strongly as you have indicated in my poor prayers, then all the more greatly can you trust in a soul so saintly, assuring yourself that by her merits you will escape every danger.'

She could not similarly endorse another offering – an electuary of dried figs, nuts, rue leaves and salt, 'held together with as much honey as was needed' – but still she prepared it for him, pressed it on him and prescribed its usage: 'You may take it every morning, before eating, in a dose about the size of a walnut, followed immediately by drinking a little Greek or other good wine, and they say it provides a marvellous defence.' Indeed, this formula carried the official endorsement of the Magistracy of Public Health.

Galileo reciprocated with periodic gifts of money and food – including meats, sweets and even a special spinach dish he cooked himself especially for her – as well as by giving her a warm quilt to replace the one she had yielded to Suor Arcangela, miscellaneous containers and glass vials for her apothecary work, and citrons that she returned to him as quickly as she could, candied to his liking. When he forgot to send a promised telescope, she reminded him to put it in the basket next time. This gentle reprimand – her only surviving mention of any scientific instrument – hints that she may have glimpsed the moons of Jupiter or the horns of Venus, despite the duties that ate up all her time.

Galileo apparently accompanied the tokens of his paternal love with frank praise for her abilities, though she deflected these compliments: 'I am confounded to hear that you save my letters', she remarked, 'and I suspect that the great love you bear me makes them seem more accomplished than they really are.'

In December 1630, the *tramontana* – the cold wind from the Apennines – beat into Tuscany with force, effectively locking Galileo indoors. Unable to visit the convent as he had wished, he sent news of his health via his new housekeeper, La Piera, whose prudence and capability impressed Suor Maria Celeste. Now she calmed herself to think that her father was well looked after, even if the attention came from a hired employee instead of her brother's family.

Suor Maria Celeste had disapproved of Vincenzio's flight but held her tongue for fear he would leave anyway, and that her intrusion would only upset him. Now she worried what might happen to her brother's empty, unguarded house in his absence. At the same time she encouraged Galileo to continue to play the generous father – 'especially in perpetuating your beneficence towards those who repay you with ingratitude, for truly this action, being so rife with difficulty, is all the more perfect and virtuous'.

In early January, the magistracy sent out nuncios and trumpeters to declare a general quarantine of forty days' duration, beginning on the 10th of the month. Sensing a decline in the epidemic's virulence, the authorities hoped to hasten it to its end by this drastic measure. The imposition of the general quarantine limited travel in or out of the city even more strictly than before, and also forbade ordinary household visits among neighbours and friends. The only permissible reasons for venturing out of one's doors now were to go to church or to buy food or medicines. Since the ordinance applied to Florence and all its surrounding communities, including Prato, Vincenzio could not have returned easily during these six weeks even if he had wanted to.

Vincenzio's new baby, Carlo, arrived on 20 January, but Galileo did not hear a timely announcement of the birth, or any word at all from his absent son and daughter-in-law regarding their welfare in this time of constant sorrow. Suor Maria Celeste's concern naturally stayed focused on the firstborn, nicknamed 'Galileino', who had been introduced to his aunts and cousin at the convent in his infancy. Now she repeatedly asked her father to bring the boy back to visit her again, as soon as it was safe to do so. 'I want you to give one more kiss', she would say in closing a letter, 'to Galileino for my love', and she sent him pine-cones to play with, expecting he could amuse himself extracting the nuts. She baked little biscuits for the other child now in Galileo's care – his grand-niece Virginia, the daughter of Vincenzio Landucci and his sickly wife – but Suor Maria Celeste could not help her father place 'La Virginia' at San Matteo.

'I am terribly disturbed by my inability to grant you satisfaction as I would have liked to do in taking custody here of La Virginia, for whom I feel such fondness, considering all the sweet relief and diversion she has been to you, Sire. However, I know that our superiors have declared themselves totally opposed to our admitting young girls, either as nuns or as charges, because the extreme poverty of our Convent, with which you are well acquainted, Sire, makes it a struggle to sustain those of us who are already here, let alone consider the addition of new mouths to feed.'

With her sister-in-law Sestilia in hiding at Montemurlo and still incommunicado after the quarantine ended, Suor Maria Celeste resumed the upkeep of Galileo's wardrobe. 'I am returning the bleached collars, which, on account of their being so frayed, refused to turn out as exquisitely as I would have wished: if you need anything else please remember that nothing in the world brings me greater joy than serving you, just as you for your part seem to devote yourself to doting on me and satisfying any request, since you provide for my every need with such solicitude.'

If her claustration posed no impediment to their emotional closeness, the distance between Bellosguardo and Arcetri now raised a formidable barrier. 'I find my thoughts stay fixed on you day and night, and many times I regret the great remove that bars me from being able to hear daily news of you, as I would so desire.'

Galileo suffered this same lonely longing. The prospect of the mule trek from his house to her convent, though it hadn't seemed so lengthy or arduous at first, of late too often held him back. He decided to move to Arcetri.

With her characteristic ingenuity and energy, Suor Maria Celeste canvassed the local property market from inside the walls of San Matteo. She discovered only a few available houses, however, because wealthy city dwellers cherished their villas in Arcetri as weekend escapes to the country, and because the working farms of the region passed down through families, as they had for generations. Moreover, some lands were deeded to the Church under a variety of legal situations, so that the question of their ownership generated considerable dispute.

'As far as I have been able to determine,' she reported of her earliest efforts, 'the priest of Monteripaldi has no jurisdiction over the villa of Signora Dianora Landi save for a single field. I understand, however, that the house was assigned as a dowry to a chapel of the Church of Santa Maria del Fiore, and this is the reason that our same Signora Dianora finds herself in litigation . . . I have also learned that Mannelli's villa is not yet taken, but is available for rent. This is a very beautiful property, and people say its air is the best in the whole region. I do not believe that you will lack the opportunity to secure it, Sire, if events turn out as well as you and I so strongly desire.'

At Eastertime, when the plague appeared to have spent itself, Galileo's prodigal son returned at last with his family and joined in the house-hunting activities, which continued all through that spring and summer. From his city home, Vincenzio could walk up to Arcetri in just a quarter of an hour to appraise prospective sites on Galileo's behalf.

'Sunday morning brought Vincenzio here to see the Perini's villa, and as Vincenzio himself will no doubt tell you, Sire, the buyer will have every advantage . . . I come to beg you not to let this opportunity slip through your fingers, for God knows when another such as this will present itself, now that we see how people who own lands in these parts cling to them.' But Villa Perini fell into another's hands.

Then, in early August, her searching turned up a property literally around the corner from the convent – even closer than her mother's house on the Ponte Corvo had been to Galileo's Via Vignali home during childhood in Padua.

'Because I do so desire the grace of your moving closer to us, Sire, I am continually trying to learn when places here in our vicinity are to be let. And now I hear anew of the availability of the villa of Signor Esaù Martellini, which lies on the Piano dei Giullari, and adjacent to us. I wanted to call it to your attention, Sire, so that you could make inquiries to see if by chance it might suit you, which I would love, hoping that with this proximity I would not be so deprived of news of you, as happens to me now, this being a situation I tolerate most unwillingly.'

The Martellini villa on the Piano dei Giullari, or field of minstrels, occupied such an ideal site atop the western slope of Arcetri that it had been dubbed 'Il Gioiello' – the jewel. Built in the 1300s by gentry who were establishing a farm near the convent, it narrowly escaped total destruction during the Siege of Florence two centuries later, when the Medici returned to power by force after one of their periodic banishments from the city. Between October 1529 and August 1530, forty thousand mostly Spanish troops, controlled by the prince of Orange and the Medici pope Clement VII, camped in the hills around the city, unwilling to enter combat but hoping to starve the Florentines into submission. The bubonic plague, attacking again at that same time, helped the army carry out its plan. Despite the lack of open hostilities, the mere presence of all those soldiers for ten straight months left the landscape a scene of devastation.

Galileo's house, Il Gioiello, in Arcetri,
where he lived from 1631 to 1642

A fresco painted in the mid-1500s in the Palazzo Vecchio, where
the triumphant Medici family resided before moving to the Pitti
Palace, shows the region of the Piano dei Giullari during the siege,
with military tents popping up like ant-hills and Il Gioiello clearly
identifiable among the neighbourhood houses.

Il Gioiello stood in ruins for several years after the siege. Then
it was refurbished and rebuilt, with thick stone walls that met at
the corners of rooms in graceful arches called lunettes, with floors
of brick laid out in herring-bone patterns, with intricate wooden
ceilings and wide windows that were shuttered, barred and set so
low they seemed to kneel into the street. Four very large rooms
and three smaller ones shared the ground floor, with a kitchen and
wine cellar below and rooms for two servants above.

What Galileo loved best about the place was the sunny garden
to the south of the house, reliably watered by the *tramontana*, and
the semi-enclosed loggia facing the courtyard near the well, where
potted fruit trees might pass the colder months in safety.

'We lament the time away from you, Sire, covetous of the
pleasure we would have drawn from this day, had we found

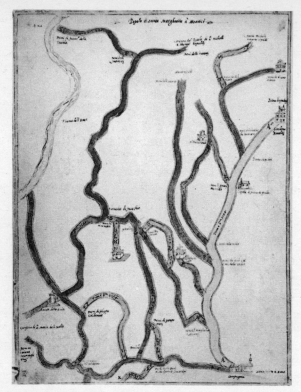

Sixteenth-century map of Arcetri; the convent of San Matteo is
in the bottom righthand corner, Il Gioiello at crossroads above

ourselves all together in each other's company. But, if it please
God, I expect that this will soon come to pass, and in the meanwhile I enjoy the hope of having you here always near us.'

The contract Galileo signed with Signor Martellini on 22
September 1631, set the rent for Il Gioiello at thirty-five *scudi* per
year – only a fraction of the hundred he had paid at Bellosguardo
– due in two equal instalments every May and November.
Approaching his seventieth year, he expected to live out the
remainder of his days in these idyllic surroundings.

From the window of the room Galileo chose for his study, he
could see the Convent of San Matteo just a stone's throw away,
downhill to the left of the vineyard.

[XX]

That I should be begged to publish such a work

BETWEEN AUGUST OF 1630, when the plague began permeating the streets of Florence, and the autumn of 1631, when Galileo settled in Arcetri, he gradually and with great difficulty broke the impasse that had stalled the publication of his *Dialogue* for another year.

Soon after Prince Cesi's sudden midsummer death in 1630, Galileo made arrangements with a new publisher and printer in Florence. In September he secured official local permission from the Florentine bishop's vicar, the Florentine inquisitor, and the grand duke's official book reviewer and court censor – just as he would have done had he never gone to Rome. But then he felt obliged to inform Father Riccardi at the Vatican of these developments. The master of the Sacred Palace had, in all fairness, invested considerable time in reading and discussing the *Dialogue* with His Holiness and had raised specific points for Galileo to treat in the text. It was only proper to inform him by letter how the demise of the Roman publisher, Prince Cesi, coupled with the disruption of communications up and down the peninsula due to the plague outbreak, necessitated publication in Tuscany.

Father Riccardi's reply let Galileo know that he had not relinquished his hold over the *Dialogue*. He still controlled its fate; in fact, he now wanted to read it again, and asked Galileo to send it back to him forthwith.

But how to send a bulky manuscript to Rome in these troubled times? The grand duke's secretary of state warned Galileo that even ordinary letters might be stopped and confiscated at checkpoints, let

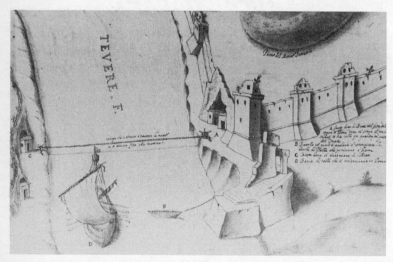

Plague barrier across the Tiber River where boats were
halted for inspection

alone whole volumes emanating from regions tainted by the epidemic.

Galileo wrote to Father Riccardi a second time, offering as a compromise to send him only the contested parts of the manuscript: the preface and the ending. Father Riccardi could change these however he saw fit – cutting, rewording, even inserting caveats 'giving these thoughts of mine the labels of chimeras, dreams, paralogisms and empty images'. As for the rest of the work, Galileo suggested it be reviewed this time within the city walls of Florence, by some authority of Father Riccardi's choosing. Father Riccardi agreed.

In November 1630, Galileo duly turned over his *Dialogue* to the new designated reviewer in Florence, Fra Giacinto Stefani, who read the work minutely. 'Indeed,' Galileo could now justly boast, 'His Paternity has stated that at more than one place in my book tears came to his eyes when seeing with how much humility and reverent submission I defer to the authority of superiors, and he acknowledges (as do all those who have read the book) that I should be begged to publish such a work.'

But still Galileo could not publish it – was most assuredly prohibited from publishing it – until Father Riccardi blessed the preface and the ending. While Galileo waited, Father Riccardi dallied. The whole winter came and went without a word from Rome. 'In the meantime the work stays in a corner', Galileo observed bitterly, 'and my life wastes away as I continue living in constant ill health.'

In March, Galileo appealed to the grand duke for help, 'so that while I am still alive I may know the outcome of my long and hard work'. Ferdinando, feeling ever kindly disposed to the ageing court philosopher who had tutored him in childhood, interceded. And Father Riccardi, who, as a native of Florence took special note of the grand duke's interest, agreed at last in late May that all but the preface and conclusion could go on press while he continued to refine those parts.

'I want to remind you', Father Riccardi wrote to the Florentine inquisitor on 24 May 1631, 'that His Holiness thinks the title and subject should not focus on the ebb and flow of the sea but absolutely on the mathematical examination of the Copernican position on the Earth's motion . . . It must also be made clear that this work is written only to show that we do know all the arguments that can be advanced for this side, and that it was not for lack of knowledge that the decree [the Edict of 1616] was issued in Rome; this should be the gist of the book's beginning and ending, which I will send from here properly revised.'

Thus the slow work of producing the large run of one thousand copies began in June. It took the entire month to set and print the first forty-eight of the *Dialogue*'s five hundred pages.

Father Riccardi pronounced his last word on the subject on 19 July, when he forwarded the preface and ending particulars to the inquisitor at Florence. 'In accordance with the order of His Holiness about Signor Galilei's book,' Father Riccardi wrote in a brief cover letter, 'I send you this beginning or preface to be placed on the first page; the author is free to change or embellish its verbal expressions as long as he keeps the substance of the content.' Father Riccardi's enclosure, originally written by Galileo, had been very lightly edited. And Galileo now changed only one word; otherwise the suggested preface and the later published version agree exactly.

Rather than dictate the precise wording of the ending, after all that had transpired, Father Riccardi simply tacked on a coda to his text for the preface. 'At the end', these instructions stated, 'one must have a peroration of the work in accordance with this preface. Signor Galilei must add the reasons pertaining to divine omnipotence which His Holiness gave him; these must quiet the intellect, even if there is no way out of the Pythagorean arguments.'

Father Riccardi knew just how compellingly Galileo had presented the 'Pythagorean arguments', as he called the Copernican view. Indeed, by the time readers turned the *Dialogue*'s last page, they might well believe that Pythagoras and Copernicus had trounced Aristotle and Ptolemy for good. But of course Galileo was not allowed to leave the matter hanging there, as though admitting the absolute truth of the Copernican opinion. Short of divine revelation, only hypothetical truth would serve.

'After an infinity of cares, finally the preface to your distinguished work has been corrected,' Tuscan ambassador Francesco Niccolini wrote to Galileo, rejoicing with him after Father Riccardi, who was his wife's cousin, had consented at last. 'The Father Master of the Sacred Palace indeed deserves to be pitied, for exactly during these days when I was spurring and bothering him, he has suffered embarrassment and very great displeasure in regard to some other works recently published, as he must have done at other times too; he barely complied with our request, and only because of the reverence he feels for the Most Serene name of His Highness our Master and for his Most Serene House.'

Certainly Galileo owed a great debt to the intervention of young Ferdinando de' Medici. He began to repay it with the special compliment of dedicating the *Dialogue* to the grand duke in a verbal bow – immediately preceding the crucial 'Preface to the Discerning Reader' that Father Riccardi had stipulated.

'These dialogues of mine revolving principally around the works of Ptolemy and Copernicus,' wrote Galileo to Ferdinando,

> it seemed to me that I should not dedicate them to anyone except Your Highness. For they set forth the teaching of these two men whom I consider the greatest minds ever to have left us such contemplations in their works; and, in order to avoid any loss of greatness, must be placed under the protection of the greatest support I know from which they can receive fame and patronage. And if those two men have shed so much light upon my understanding that this work of mine can in large part be called theirs, it may properly be said also to belong to Your Highness, whose liberal munificence has not only given me leisure and peace for writing, but whose effective assistance, never tired of favouring me, is the means by which it finally reaches publication.

Meanwhile, the exhaustive process of printing the *Dialogue* wore on. By mid-August, when one-third of the pages had piled up, Galileo told friends in Italy and France that he hoped to see the rest finished by November. But it took even longer, so that a total of nine months passed from the start of printing to the book's completion in February 1632. Its wordy title filled a page:

Dialogue
of
Galileo Galilei, Lyncean
Special Mathematician of the University of Pisa
And Philosopher and Chief Mathematician
of the Most Serene
Grand Duke of Tuscany.
Where, in the meetings of four days, there is discussion
concerning the two
Chief Systems of the World,
Ptolemaic and Copernican,
Propounding inconclusively the philosophical and physical reasons
as much for one side as for the other.

No written words of encouragement from Suor Maria Celeste sped Galileo through this last leg of publication – simply because father and daughter now lived in such proximity that they found no reason to write. A short walk took him from his front door to her parlour grille in minutes, and if he was too busy or too troubled by his pains, he could send La Piera with news and a basket of something. After Suor Maria Celeste addressed her last letter to Galileo at Bellosguardo on 30 August 1631, she may have imagined she need never write to her father again. But his move did not end their correspondence. It merely introduced a pause that lasted almost one and a half years – until early in 1633, when the shock wave initiated by the *Dialogue*'s publication boomeranged and shattered Galileo's peace in Arcetri.

At first, everything augured well for the book, which met with immense and immediate success. Galileo presented the first bound

Frontispiece of Galileo's *Dialogue*; the three figures represent,
from left to right, Aristotle, Ptolemy and Copernicus

copy to the grand duke at the Pitti Palace on 22 February 1632. In
Florence the book sold out as quickly as it entered the shops.
Galileo also sent copies to friends in other cities, such as Bologna,
where a fellow mathematician commented, 'Wherever I begin, I
can't put it down.'

The copies destined for Rome, however, were held up until
May on the advice of Ambassador Niccolini, who apologised that
current Roman quarantine regulations required all shipments of
imported books to be dismantled and fumigated – and no one
wanted to see the *Dialogue* subjected to such treatment. Galileo

got around this obstacle by sending several presentation copies into Rome via the luggage of a travelling friend, who distributed them to various luminaries including Francesco Cardinal Barberini. Galileo's long-time confrère, Benedetto Castelli, now 'Father Mathematician of His Holiness', read one of these copies.

'I still have it by me,' Castelli wrote to Galileo on 29 May 1632, 'having read it from cover to cover to my infinite amazement and delight; and I read parts of it to friends of good taste to their marvel and always more to my delight, more to my amazement, and with always more profit to myself.'

A young and as yet unknown student of Castelli's named Evangelista Torricelli* wrote to Galileo in the summer of 1632 to say he had been converted to Copernicanism by the *Dialogue*. The Jesuit fathers with whom he had formerly studied, he told his new idol, had also taken great pleasure in the book, though naturally they could not corroborate the opinions of Copernicus.

Some Jesuit astronomers, however, especially Father Christopher Scheiner, the 'Apelles' who claimed to have discovered sunspots before Galileo, reacted violently to the *Dialogue*. Scheiner's own latest book, the long-delayed *Rosa Ursina*, which finally appeared in April 1631, had lambasted Galileo with offensive language. Now Scheiner was living in Rome, having learned to speak Italian, and he harangued Father Riccardi to have the *Dialogue* banned. On top of the anger he had apparently stewed in his breast since the sunspot debate two decades earlier, Scheiner felt newly annoyed by what he inferred to be a fresh personal slander against him in Galileo's book.

Soon the *Dialogue* provoked Pope Urban's ire as well. It came to his attention at a most inopportune moment, when his profligate spending on war efforts was well on its way to doubling the papal debt, and when his fears of Spanish intrigue against him had reached new heights of paranoia. At a private consistory Urban had held with the cardinals on 8 March 1632 the Vatican ambassador to Spain, Gaspare Cardinal Borgia, had openly censured the

---

* Torricelli (1608–47) is remembered today as the inventor, in 1643, of the barometer.

pontiff's failure to back King Philip IV in the Thirty Years' War against the German Protestants. The pope's behaviour, Cardinal Borgia charged, evinced his inability to defend the Church – even his unwillingness to do so. Hasty efforts by Urban's family cardinals to silence the Spanish sympathiser almost came to physical blows before the Swiss Guard entered the chamber to restore order.

Fearing poison, Urban secluded himself at Castel Gandolfo, a lakeside papal holiday retreat thirteen miles southeast of Rome. He suspected Spanish-controlled military manoeuvres in Naples of being aimed at him, and he imagined that the grand duke of Tuscany, any day now, would sail his navy into the papal ports of Ostia and Civitavecchia, in retribution for Urban's appropriating Medici holdings in Urbino.

Although Florentine himself, Urban had encroached on Medici property early in his pontificate, in 1624, by laying unlawful claim to land Ferdinando was due to inherit from the elderly and infirm Francesco della Rovere, the duke of Urbino. Pope Urban decided that the duke's death would leave Urbino a vacant fief, which he could annex to the states of the Church. But Ferdinando's aunt Caterina de' Medici, formerly the duchess of Urbino, had long ago left the territory to Ferdinando's family in her will. Also Ferdinando's bride-to-be, to whom he had been betrothed when he was twelve and she an infant in arms, was Vittoria della Rovere, the granddaughter and only heir of the ageing duke. The primary purpose of the couple's long-standing engagement had been to secure the Duchy of Urbino for the House of Medici. These particulars, however, did not stop Urban from marching the papal troops into Urbino, poised to take possession. After Francesco della Rovere finally died in 1631, Ferdinando and Vittoria (still a child living in the Florentine convent of the Crocetta) had lost the land to Pope Urban.

When Galileo's book arrived in Rome in the summer of 1632, Urban could take no time to read it. Anonymous advisers judged it for him, however, as an egregious insult. Galileo's enemies in Rome, whose number was legion, saw the *Dialogue* as a scandalous glorification of Copernicus. And the pope, already loudly accused of flagging Catholic zeal on the battle-fronts of Europe, could not allow a new affront to go unpunished.

In August His Holiness, stung by inflammatory remarks insisting Galileo had played him for a fool by allowing Simplicio to espouse Urban's philosophy, convened a three-man commission to re-examine the text of the *Dialogue*. 'We think that Galileo may have overstepped his instructions by asserting absolutely the Earth's motion and the Sun's immobility, thus deviating from hypothesis,' these commissioners said in their September report to the pope. 'One must now consider how to proceed, both against the person and concerning the printed book.'

Ambassador Niccolini and the grand duke's secretary of state, who kept up a flurry of secret diplomatic correspondence during these developments, agreed morosely that 'the sky seemed about to fall'. 'I feel the Pope could not have a worse disposition towards our poor Signor Galilei,' the ambassador wrote on 5 September, recounting the results of a papal audience conducted 'in a very emotional atmosphere', during which Urban had 'exploded into great anger' and then railed on 'with that same outburst of rage'.

'When His Holiness gets something into his head, that is the end of the matter,' wrote Niccolini, speaking from unpleasant experience, 'especially if one is opposing, threatening or defying him, since then he hardens and shows no respect to anyone . . . This is really going to be a troublesome affair.'

Before the end of September, an official order reached the inquisitor at Florence, announcing that the *Dialogue* could no longer be sold (though it was already sold out) and demanding that the author appear before the Holy Office of the Inquisition during the month of October.

Galileo applied for leniency to Francesco Cardinal Barberini, his most powerful friend, although these harsh commands had actually issued from the pope's brother Antonio, called Cardinal Sant' Onofrio. Would Urban VIII please excuse the aged, unwell Galileo from travelling to Rome – especially now that plague was breaking out again in Florence? And, given the fact that the *Dialogue* had gone through proper channels to receive official approval from all the relevant authorities, couldn't Galileo respond in writing to any objection now raised against it?

No. And no. The most the angry pontiff would concede was that Galileo might travel to Rome in comfort and at his own pace, but come he must. And soon. Already, delays caused by his jockeying for appeals had swallowed the whole month of October, and Galileo would lose at least another twenty to forty days quarantined at some midway point – Siena perhaps – before being allowed into Rome.

November found Galileo sick in bed, however, too ill to go anywhere. The pope fumed, especially as the illness wore on into December, when the Florentine inquisitor paid a house call on Galileo at Arcetri. There, a panel of three prominent doctors, including Galileo's friend and personal physician Giovanni Ronconi, signed an affidavit on 17 December, listing a long series of ailments: intermittent pulse indicating the general weakness of declining years, frequent vertigo, hypochondriacal melancholy, weakness of the stomach, diverse pains throughout the body, serious hernia with rupture of the peritoneum. In short, moving him would put his life in jeopardy.

The inquisitors dismissed the report in disbelief. Galileo could come to Rome of his own free will, they decreed, or he could be arrested and dragged there in irons. Grand Duke Ferdinando, powerless to oppose the will of the pope in this case, eased Galileo's way by once more lending him a litter and a servant to attend him on his journey.

Fully cognizant of the gravity of his circumstances, the sixty-eight-year-old Galileo made out his will and wrote a long, rueful letter to his friend Elia Diodati in Paris shortly before leaving Arcetri. 'I am just now going to Rome,' this letter of 15 January 1633 said in part,

> whither I have been summoned by the Holy Office, which has already prohibited the circulation of my *Dialogue*. I hear

from well-informed parties that the Jesuit Fathers have insinuated in the highest quarters that my book is more execrable and injurious to the Church than the writings of Luther and Calvin. And all this although, in order to obtain the imprimatur, I went in person to Rome and submitted the manuscript to the Master of the Sacred Palace, who looked through it most carefully, altering, adding and omitting, and, even after he had given it the imprimatur, ordered that it should be examined again at Florence. The reviser here, finding nothing else to alter, in order to show that he had gone through it carefully, contented himself with substituting some words for others, as, for instance, in several places, 'Universe' for 'Nature', 'quality' for 'attribute', 'sublime spirit' for 'divine spirit', excusing himself to me for this by saying he foresaw that I should have to do with fierce foes and bitter persecutors, as has indeed come to pass.

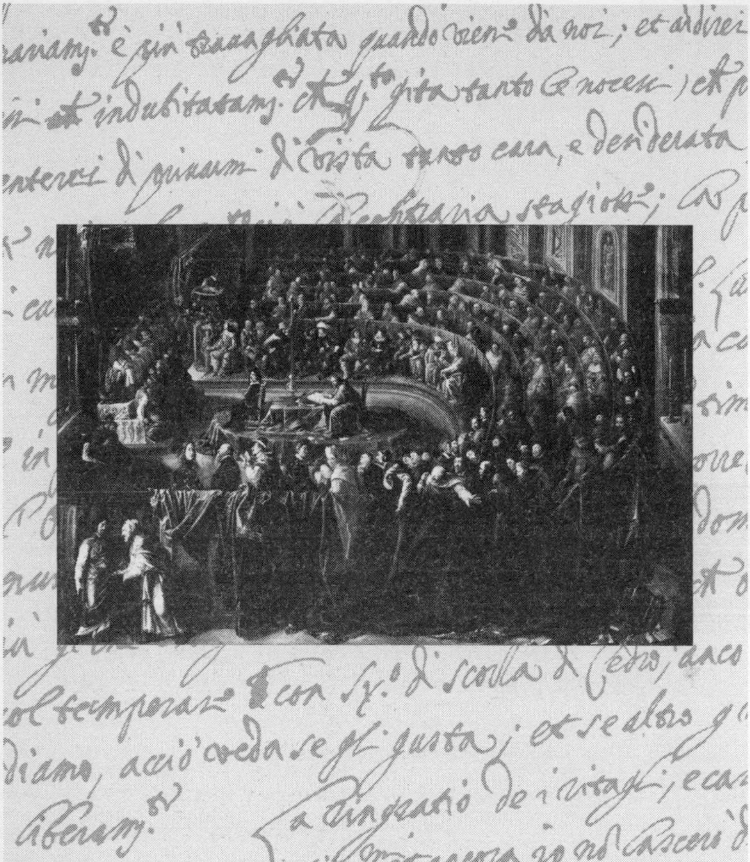

PART FOUR

Care of the Tuscan Embassy, Villa Medici, Rome

[XXI]

How anxiously I live, awaiting word from you

THERE WAS ONLY ONE trial of Galileo, although legends – even
experts and encyclopedias – often speak of two, erroneously
counting Galileo's 1616 encounter with Cardinal Bellarmino as a
preliminary trial, leading up to the second, more sustained inter-
rogation of 1633 that left Galileo kneeling before his inquisitors, or
in a dungeon by some accounts, or even in chains.

There was only one trial of Galileo, and yet it seems there were
a thousand – the suppression of science by religion, the defence of

individualism against authority, the clash between revolutionary and establishment, the challenge of radical new discoveries to ancient beliefs, the struggle against intolerance for freedom of thought and freedom of speech. No other process in the annals of canon or common law has ricocheted through history with more meanings, more consequences, more conjecture, more regrets.

The confusion over Galileo's trial – whether one or two actually took place, and when – derives from the abstruse nature of the trial itself. There was only one trial of Galileo, in the spring of 1633, but at least half of the evidence and most of the testimony involved contested events of 1616.

The trial testimony, which survives thanks to careful recording at the time, accentuates the alienation between accuser and accused by its very choice of language: the transcript summarises the prosecutor's inquiries in Latin in the third person, so that the questions on paper assume a quasi-historical cast ('By what means and how long ago did he come to Rome?'), while the defendant's responses ring small and meek in first-person Italian ('I arrived in Rome the first Sunday of Lent and I came in a litter'). Thus, although the recitative is marked Q and A throughout, the two parts refuse to blend. The text of the courtroom drama continually jars the reader by presenting two speakers staged as though to engage one another, while each pursues his own stream of consciousness.

After Galileo left Arcetri for Rome on 20 January 1633, he spent nearly two weeks travelling and then passed another two weeks detained near Acquapendente in quarantine – in uncomfortable quarters, with nothing to eat but bread and eggs with wine – so that he entered the Holy City on Sunday night, 13 February.

Urban could have had him gaoled immediately, but instead, in a respectful gesture to Grand Duke Ferdinando and in deference to Galileo's frail health, the pope allowed him to stay at the Tuscan embassy, next door to the Church of the Trinità del Monte, where he had lodged comfortably during previous visits. His hosts, Francesco and Caterina Niccolini, welcomed Galileo as their honoured guest and tried to mitigate the gravity of his circumstances with the warmth of their hospitality.

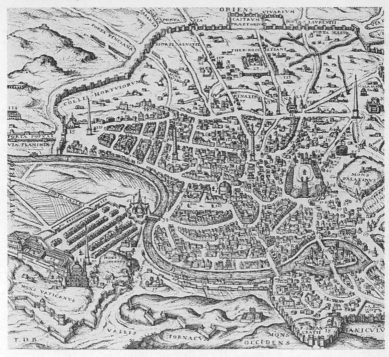

Rome in 1596

Ambassador Niccolini had been intimately involved in the prelude to Galileo's current predicament, having pleaded his case with Father Riccardi, with Francesco Cardinal Barberini, and on several occasions with Pope Urban at the peak of his spleen. The ambassador had succeeded in learning as much as could be expected, given the fact, as he explained to his superiors in Tuscany, that 'we are dealing with the Congregation of the Holy Office, whose goings-on are so secret and none of whose members opens his mouth because of the censures that are in force'.

Now, with Galileo in his house awaiting God only knew what fate, Niccolini continued visiting various cardinals and trying to help his old friend in every way he could imagine. Galileo did not go along on these excursions but stayed at the embassy, under orders from Cardinal Barberini to seclude himself for his own

The Villa Medici in Rome

protection. The only person who called on Galileo was a certain Monsignor Lodovico Serristori, a consultant to the Holy Office.

'The latter has come twice,' Niccolini observed at the end of Galileo's first week in residence, 'claiming to be acting on his own and to want to visit; but he has always mentioned the trial and discussed various details, and so I believe one may be certain he has been sent to hear what Signor Galilei says, what his attitude is, and how he defends himself, so that they can then decide what to do and how to proceed. These visits seem to have comforted this good old man by encouraging him and giving him the impression that they are interested in his case and in what decisions are being taken. Nevertheless, sometimes this persecution seems very strange to him.'

Niccolini, whose ingenuous, meticulous letters to the Tuscan secretary of state over the next two months constitute a summary of the pre-trial hearings, told his houseguest everything he knew. From the files of the Holy Office, an ominous document had surfaced that some considered sufficient to ruin Galileo. The paper dated from his visit to Rome of December 1615 until June 1616 – long before Ferdinando had become grand duke, before Niccolini had been named ambassador, before Urban was elected pope.

These old notes from Galileo's Inquisition dossier, Niccolini explained, showed that Galileo had been officially warned not to discuss Copernicus, ever, in any way at all. And so, when Galileo had come to Urban in 1624, testing the feasibility of treating Copernican theory hypothetically in a new book, he had in fact been flouting this ruling. Worse, it now appeared he had intentionally duped the trusting Urban by not having had the decency to tell him such a ruling even existed. No wonder the pope was furious.

Galileo felt certain these notes Niccolini mentioned must refer to the warning Cardinal Bellarmino had given him, hat in hand, just before the pronouncement of the edict. But the late cardinal's warning had *not* been so strictly explicit as Niccolini's information now seemed to indicate. It had left leeway for hypothetical discussion. Freedom to discuss the topic hypothetically was all Galileo had asked of Urban, and all that he had done. Surely the whole unfortunate misunderstanding could be resolved once his side was heard.

But Niccolini feared that His Holiness and the Holy Office, having made a great show of dragging Galileo to their doorstep, would not admit to having blundered by arresting an innocent man.

Galileo, after being harried to Rome by repeated threats, now frittered away weeks in the Tuscan embassy, waiting to be called for questioning. The empty hours made him hungry for news of home. Departing Arcetri, uncertain how long he'd be gone, he had offered the use of his villa to Francesco Rondinelli, librarian to Grand Duke Ferdinando and chronicler of the current plague situation.* Galileo expected his housekeeper and servant boy, La Piera and Giuseppe, to tend their chores as usual, and he deputised Suor Maria Celeste to assume executive control, from inside the convent, of all his personal and household affairs.

---

* Rondinelli's book, *Report on the Contagion in Florence During the Years 1630 and 1633*, was published in 1634, the year after the epidemic ended in a miracle.

Most Illustrious and Beloved Lord Father

YOUR LETTER WRITTEN ON the 10th of February was delivered to me on the 22nd of the same month, and by now I assume you must have received another letter of mine, Sire, along with one from our Father Confessor, and through these you will have learned some of the details you wanted to know; and seeing that still no letters have come giving us definite news of your arrival in Rome (and you can imagine, Sire, with what eagerness I in particular anticipate those letters), I return to write to you again, so that you may know how anxiously I live, while awaiting word from you, and also to send you the enclosed legal notice, which was delivered to your house, 4 or 5 days ago, by a young man, and accepted by Signor Francesco Rondinelli, who, in giving it to me, advised me that it must be paid, without waiting for some more offensive insult from the creditor, telling me that one could not disobey such an order in any manner, and offering to handle the matter himself. This morning I gave him the 6 *scudi,* which he did not want to pay to Vincenzio* but chose to deposit the money with the magistrate until you have told him, Sire, what you want him to do. Signor Francesco is indeed a most pleasant and discreet person, and he never stops declaiming his gratefulness to you, Sire, for allowing him the use of your house. I heard from La Piera that he treats her and Giuseppe with great kindness, even in regard to their food; and I provide for the rest of their needs, Sire, according to your directions. The boy tells me that this Easter he will need shoes and stockings, which I plan to

---

* Her cousin Vincenzio Landucci had apparently found some pretext for filing a lawsuit against Galileo.

knit for him out of thick, coarse cotton or else from fine wool. La Piera maintains that you have often spoken to her about ordering a bale of linen, on which account I refrained from buying the small amount I would need to begin weaving the thick cloth for your kitchen, as I had meant to do, Sire, and I will not make the purchase unless I hear otherwise from you.

The vines in the garden will take nicely now that the Moon is right, at the hands of Giuseppe's father, who they say is capable enough, and also Signor Rondinelli will lend his help. The lettuce I hear is quite lovely, and I have entrusted Giuseppe to take it to be sold at market before it spoils. From the sale of 70 bitter oranges came 4 *lire*, a very respectable price, from what I understand, as that fruit has few uses: Portuguese oranges are selling for 14 *crazie* per 100 and you had 200 that were sold.

As for that barrel of newly tapped wine you left, Sire, Signor Rondinelli takes a little for himself every evening, and meanwhile he makes improvements to the wine, which he says is coming along extremely well. What little of the old wine that was left I had decanted into flasks, and told La Piera that she and Giuseppe could drink it when they had finished their small cask, since we of late have had reasonably good wine from the convent, and, being in good health, have hardly taken a drop.

I continue to give one *giulio* every Saturday to La Brigida, and I truly consider this an act of charity well deserved, as she is so exceedingly needy and such a very good girl.

Suor Luisa, God bless her, fares somewhat better, and is still purging, and having understood from your last letter, Sire, how concerned you were over her illness out of your regard for her, she thanks you with all her heart; and while you declare yourself united with me in loving her, Sire, she on the other hand claims to be the paragon of this emotion, nor do I mind granting her that honour, since her affection stems from the same source as yours, and it is myself; wherefore I take pride in and prize this most delicious contest of love, and the more clearly I perceive the greatness of that love you both bear me, the more bountiful it grows for being mutually exchanged between the very two persons I love and revere above everyone and everything in this life.

Tomorrow will be 13 days since the death of our Suor Virginia Canigiani, who was already gravely ill when I last wrote to you, Sire, and since then a malevolent fever has stricken Suor Maria Grazia del Pace, the eldest of the three nuns who play the organ, and teacher of the Squarcialupis, a truly tranquil and good nun; and since the doctor has already given her up for dead, we are all beside ourselves, grieving over our loss. This is everything I need to tell you for the moment, and as soon as I receive your letters (which must surely have arrived at Pisa by now where the Bocchineri gentlemen are) I will write again. Meanwhile I send you the greetings of my heart together with our usual friends, and particularly Suor Arcangela, Signor Rondinelli and Doctor Ronconi, who begs me for news of you every time he comes here. May the Lord God bless you and keep you happy always.

FROM SAN MATTEO, THE 26TH DAY OF FEBRUARY 1633.
Most affectionate daughter,

S. M. Celeste Galilei

Signor Rondinelli, having this very moment returned from Florence, tells me he spoke to the Chancellor of the Advisers and learned that the 6 *scudi* must be paid to Vincenzio Landucci and not be deposited, and this will be done; I submitted to this decision reluctantly, not having had your instructions on the matter.

Suor Maria Celeste's signature on this letter included, for the first time, her surname, as though to assert the ties that united them despite the physical separation and the precarious nature of Galileo's situation. Previously she had twice tacked on the initial G, but never written out the last name in full.

The date on this letter, the 26 of February, marked an anniversary that Galileo might well have remembered while he waited to be called before the Inquisition. For on that same date, seventeen years previously, Cardinal Bellarmino had brought Galileo to his palace and instructed him to abandon his belief in the Copernican doctrine. There had come a similar warning almost simultaneously, from the commissary general of the Holy Office, who had appeared with some other men in the cardinal's rooms that day. Soon afterwards the edict had been issued, and then the malicious gossipers had begun nattering how Galileo had abjured in the cardinal's hand – until he had been forced to seek a statement to the contrary from the cardinal himself.

Galileo had kept Cardinal Bellarmino's letter all these years and brought it along with him to Rome, together with copies of other correspondence, just in case.

All of February having passed without Galileo's being called to appear before the inquisitors, Suor Maria Celeste let herself believe the looming ordeal might be reduced to a tranquil interlude that would ultimately redound to Galileo's greater recognition and renown. 'I rejoice, and ever again I thank blessed God,' she wrote on 5 March, 'hearing that your affairs thus far proceed with such tranquillity and silence, which bodes well for a happy and prosperous outcome, as I have always hoped would come with divine help and by the intercession of the most holy Blessed Virgin.' So far, nothing had happened to Galileo in Rome save for his birthday's passing uneventfully. Now he was sixty-nine – although the references to him throughout the trial, including his personal declarations, stated his age as seventy years.

'About Signor Galilei I cannot report to Your Most Illustrious Lordship anything more than what I wrote in my past letters,' Ambassador Niccolini stated in his dispatch of 6 March, 'except that I am trying to arrange, if possible, that he be allowed occasionally to go into the garden of the Trinità in order to be able to exercise a little; for it is very harmful to remain always inside the house. However, so far I have not received any answer, nor do I know what we can hope on the matter.'

The grand duke tried at this point to help Galileo from afar by sending letters of recommendation to a couple of the cardinal inquisitors, requesting that they favour his dear employee. Ferdinando pursued this course despite Urban's warning him, through intermediaries, not to interfere, on the grounds that he could not come out of the affair honourably: what sort of tribute did the dedication of the *Dialogue* offer the grand duke anyway? Was not the real duty of any Christian prince to shield Catholicism from danger? Just as Urban had felt compelled to ban many a book dedicated to himself, in order to protect the Church, so he said Ferdinando should follow suit and cut himself loose from Galileo.

Instead of pulling back, however, Ferdinando redoubled his efforts. On Niccolini's advice, he wrote additional letters to all the other cardinals of the Holy Office, lest one or another of the ten inquisitors feel slighted by his oversight.

Suor Maria Celeste continued to write Galileo at least one long report every Saturday, in which she attempted 'to fit all the things I would chatter to you about over a week's time'. To allay the pain of his absence, she kept herself busier than ever – or, in her words, she performed 'the office of Martha', the patron saint of cooks and housekeepers, 'all day long . . . without a single intermission'.

Suor Maria Celeste wrote also to Ambassadress Caterina Niccolini, who had forged such a bond with the sisters of San Matteo by this point, through her many demonstrations of generosity, that she spoke of hoping to attend a religious drama at the convent.

'[Her] visit, should Suor Arcangela and I be so fortunate as to receive one,' Suor Maria Celeste confided to Galileo on 12 March, 'would certainly be a noteworthy honour and as welcome to us as you will have to imagine yourself, Sire, for I know not how to express it. As for allowing her to view a play, I am speechless, because it would have to be rehearsed in time for her arrival, while I honestly believe, since she has evinced this desire to hear us perform, Sire, that we would be safer leaving her believing in the talent she assumes us to have on the strength of your say-so.'

At the same time in mid-March, Ambassador Niccolini approached the pope once more, begging him to expedite the trial and let Galileo go home without being brought before the Inquisition. 'I reiterated that his old age, ill health and readiness to submit to any censure might render him worthy of such favour,' Niccolini wrote of this attempt, 'but His Holiness again said he thinks there is no way out, and may God forgive Signor Galilei for having meddled with these subjects.'

[XXII]

In the chambers of the Holy Office of the Inquisition

ON TUESDAY, 12 APRIL 1633, after Galileo had waited two months at the Tuscan embassy, the commissary general of the Holy Office of the Inquisition finally brought him in for questioning. Although several well-known paintings of Galileo standing before the Inquisition portray him ringed by large numbers of churchmen, he gave his actual testimony to just two officials and a secretary. The ten cardinals who were to serve as judges and jury did not attend this stage of the proceedings, which they could read later at their leisure, or be briefed about at one of their regular Wednesday morning meetings.

Most Beloved Lord Father

SIGNOR GERI [Bocchineri – Sestilia's brother and the grand duke's private secretary] informed me of the conditions imposed on you on account of your affair, Sire, that alas you are detained in the chambers of the Holy Office; on the one hand, this gives me great distress, convinced as I am that you find yourself with scant peace of mind, and perhaps also deprived of all bodily comforts: on the other hand, considering the need for events to reach this stage, in order for the authorities to dismiss you, as well as the kindliness with which everyone there has treated you up till now, and above all the justice of the cause and your innocence in this instance, I console myself and cling to the expectation of a happy and prosperous triumph, with the help of blessed God, to Whom my heart never ceases to cry out, commending you with all the love and trust it contains.

The only thing for you to do now is to guard your good spirits, taking care not to jeopardise your health with excessive worry, but to direct your thoughts and hopes to God, Who, like a tender, loving father, never abandons those who confide in Him and appeal to Him for help in time of need. Dearest lord father, I wanted to write to you now, to tell you I partake in your torments, so as to make them lighter for you to bear: I have given no hint of these difficulties to anyone else, wanting to keep the unpleasant news to myself, and to speak to the others only of your pleasures and satisfactions. Thus we are all awaiting your return, eager to enjoy your conversation again with delight.

And who knows, Sire, if while I sit writing, you may not already find yourself released from your predicament and free of all concerns? Thus may it please the Lord, Who must be the One to console you, and in Whose care I leave you.

FROM SAN MATTEO, THE 20TH DAY OF APRIL 1633.

Most affectionate daughter,

S. M. Celeste

An anxious daughter penned these words in a very controlled hand, much smaller than her usual script. But as optimistically as she might have hoped for the crisis to reach a speedy conclusion while a few letters crossed in the mail, the ordeal of her father's trial had only just begun. Its unfolding is preserved as follows in the testimony recorded verbatim at the time.

> Summoned, there appeared personally in Rome at the Palace of the Holy Office, in the usual quarters of the Reverend Father Commissary, fully in the presence of the Reverend Father Fra Vincenzo Maculano da Firenzuola, Commissary General, assisted by Lord Carlo Sinceri, Prosecutor of the Holy Office, etc. Galileo, son of the late Vincenzio Galilei, Florentine, seventy years of age, who, sworn to testify the truth, was asked by the Fathers the following:
>
> *Q:* By what means and how long ago did he come to Rome.
>
> *A:* I arrived in Rome the first Sunday of Lent, and I came in a litter.
>
> *Q:* Whether he came of his own accord, or was called, or was ordered by someone to come to Rome, and by whom.
>
> *A:* In Florence the Father Inquisitor ordered me to come to Rome and present myself to the Holy Office.

Q: Whether he knows or can guess the reason that this order was given to him.

A: I imagine that the cause of my having been ordered to come before the Holy Office is to give an account of my recently printed book; and I suppose this because of the order given to the printer and to myself, a few days before I was ordered to come to Rome, not to issue any more of those books, and similarly because the printer was ordered by the Father Inquisitor to send the original manuscript of my book to the Holy Office in Rome.

Q: That he explain what is in the book he imagines was the reason for the order that he come to the city.

A: It is a book written in dialogue, and it treats of the constitution of the world, or rather, of the two chief systems, that is, the arrangements of the heavens and of the elements.

Q: Whether, if he were shown the said book, he would recognise it as his.

A: I hope so; I hope that if it is shown to me I shall recognise it.

And there was shown to him a book printed at Florence in the year 1632, with the title *Dialogue of Galileo Galilei Lyncean* etc. [Exhibit *A*]; and when he had looked at it and inspected it, he said: 'I know this book very well, and it is one of those printed in Florence, and I acknowledge it as mine and composed by me.'

Q: Whether he likewise acknowledges each and every word contained in the said book as his.

A: I know this book shown to me, for it is one of those printed in Florence; and I acknowledge all it contains as having been written by me.

Q: When and where he composed the said book, and how long it took him.

A: As to the place, I composed it at Florence, beginning ten or twelve years ago; and I was occupied on it about six or eight years, though not continuously.

Q: Whether he was in Rome another time, particularly in the year 1616, and for what occasion.

A: I was in Rome in 1616, and afterwards I was here in the second year of the pontificate of His Holiness Urban VIII, and lastly I was here three years ago, on the occasion of my wish to have my book printed. The occasion for my being in Rome in the year 1616 was that, hearing questions raised about the opinion of Nicolaus Copernicus concerning the motion of the Earth and stability of the Sun and the order of the celestial spheres, in order to assure myself against holding any but holy and Catholic opinions, I came to hear what was proper to hold concerning this matter.

Q: Whether he came because he was summoned, and if so, for what reason he was summoned, and where and with whom he discussed the said matter.

A: In 1616 I came to Rome of my own accord, without being summoned, and for the reason I told you. In Rome I treated of this business with some Cardinals who governed the Holy Office at that time, in particular with Cardinals Bellarmino, Aracoeli, San Eusebio, Bonsi and d'Ascoli.

Q: What specifically he discussed with the said cardinals.

A: The occasion for discussing with these cardinals was that they wished to be informed of the doctrine of Copernicus, his book being very difficult to understand for those outside the mathematical and astronomical profession. In particular they wanted to know the arrangement of the celestial orbs under the Copernican hypothesis, how he places the Sun at the centre of the planets' orbits, how around the Sun he places next the orbit of Mercury, around the latter that of Venus, then the Moon around the Earth, and around this Mars, Jupiter and Saturn; and in regard to motion, he makes the Sun stationary at the centre and the Earth turn on itself and around the Sun, that is, on itself with the diurnal motion and around the Sun with the annual motion.

Q: Since, as he says he came to Rome to be able to have the truth about the said matter, let him state also what was the outcome of this business.

A: Concerning the controversy that went on about the said opinion of the stability of the Sun and motion of the Earth, it was determined by the Holy Congregation of the Index that this opinion, taken absolutely, is repugnant to Holy Scripture, and it is to be admitted only *ex suppositione*, the way in which Copernicus takes it.

Q: Whether he was then notified of the said decision, and by whom.

A: I was indeed notified of the said decision of the Congregation of the Index, and I was notified by Lord Cardinal Bellarmino.

Q: Let him state what the Most Eminent Bellarmino told him about the said decision, whether he said anything else about the matter, and if so what.

A: Lord Cardinal Bellarmino informed me that the said opinion of Copernicus could be held hypothetically, as Copernicus himself had held it. His Eminence knew that I held it hypothetically, namely in the way Copernicus held it, as you can see from an answer by the same Lord Cardinal to a letter of Father Master Paolo Antonio Foscarini, Provincial of the Carmelites; I have a copy of this, and in it one finds these words: 'I say that it seems to me that Your Reverence and Signor Galilei are proceeding prudently by limiting yourselves to speaking hypothetically and not absolutely.' This letter by the said Lord Cardinal is dated 12 April 1615. Moreover, he told me that otherwise, namely taken absolutely, the opinion could be neither held nor defended.

Q: What decision was made and then notified to him in the month of February 1616.

A: In the month of February 1616, Lord Cardinal Bellarmino told me that since the opinion of Copernicus, taken absolutely, contradicted Holy Scripture, it could not be held or defended, but that it might be taken and used hypothetically. In conformity with this I keep a certificate by Lord Cardinal Bellarmino himself, made in the month of May, on the 26th, 1616, in which he says that the opinion of Copernicus cannot be held or defended, being against the Holy Scripture. I present a copy of this certificate, and here it is.

And he exhibited a sheet of paper written on one side, about twelve lines, beginning 'We, Roberto Cardinal Bellarmino, having' and ending 'This 26th day of May, 1616', which was accepted as evidence and marked with the letter B. He then added: 'The original of this affidavit I have with me in Rome, and it is entirely written in the hand of Cardinal Bellarmino.'

Q: Whether, when he was notified of the above-mentioned matters, there were any other persons present, and who they were.

A: When Lord Cardinal Bellarmino told me what I have said about the opinion of Copernicus there were some Dominican Fathers present; but I did not know them, nor have I seen them since.

Q: The said Fathers being present at that time, whether they or anyone else gave him an injunction of any kind about the same subject, and if so what.

A: As I recall it, the affair came about in this manner: one morning Lord Cardinal Bellarmino sent for me, and he told me a certain particular, which I should like to speak to the ear of His Holiness before that of any one else; but in the end he told me that the opinion of Copernicus could not be held or defended, being contrary to Holy Scripture. As to those Dominican Fathers, I do not remember whether they were there first, or came afterwards; nor do I recall whether they were present when the Cardinal told me that the said opinion could not be held. And it may be that some precept was made to me that I might not hold or defend the said opinion, but I have no memory of it, because this was many years ago.

Q: Whether, if one were to read to him what he was then told and ordered with injunction, he would recall that.

A: I do not remember that I was told anything else, nor can I know whether I should recall what was then said to me even if it were read to me; and I say freely what I do recall, because I claim not to have contravened in any way the precept, that is, not to have held or defended the said opinion of the motion of the Earth and stability of the Sun on any account.

And having been told that the said injunction, given to him then in the presence of witnesses, states that he cannot in any way whatever hold, defend, or teach the said opinion, he was asked whether he remembers how and by whom he was so ordered.

The interrogators were referring now to the minutes of the Holy Office for the year 1616, which contained numerous entries that mentioned Galileo by name, though he had not given any deposition in the chambers himself at that time. On 25 February 1616, for example, a brief entry noted: 'His Holiness [Pope Paul V] ordered the Most Illustrious Lord Cardinal Bellarmino to summon before him the said Galileo and admonish him to abandon the said opinion; and in case of his refusal to obey, the Father Commissary, in the presence of a notary and witnesses, is to issue him an injunction to abstain altogether from teaching or defending this doctrine and opinion and even from discussing it; and further, if he should not acquiesce, he is to be imprisoned.'

Following on the same page in the Inquisition records, the next entry is dated 26 February:

> In the Palace and residence of Cardinal Bellarmino, Galileo
> being called and being in the presence of the Cardinal and of
> the Reverend Father Michelangelo Seghizzi of Lodi, of the
> Order of Preachers, Commissary General of the Holy Office,
> the Cardinal admonished the said Galileo of the error of the
> above-mentioned opinion and warned him to abandon it; and
> immediately and without delay, the said Cardinal being still
> present, the said Commissary gave Galileo a precept and
> ordered him in the name of His Holiness the Pope and the
> whole body of the Holy Office to the effect that the said
> opinion that the Sun is the centre of the universe and the
> Earth moves must be entirely abandoned, nor might he from
> then on in any way hold, teach, or defend it by word or
> in writing; otherwise the Holy Office would proceed against
> him.

The hastily added warning from the former commissary general, which may have struck Galileo as merely a rehash of Bellarmino's words in the bustle of that February morning, had thus been preserved in the Inquisition files, in the most unyielding terms: 'Nor might he from then on in any way hold, teach, or defend it by word or in writing.'

A separate communication of possible significance had apparently transpired in private between Cardinal Bellarmino and Galileo that same morning years earlier, regarding the 'certain particular' the cardinal had told him, and that Galileo testified he would not now divulge to anyone unless he could whisper it first in 'the ear of His Holiness'. Guesses abound as to the content of that secret message, which may – to cite just one possibility – have mentioned Maffeo (then Cardinal) Barberini's efforts to protect Copernicanism from being branded 'heresy'. But the prosecutor did not pursue this line of questioning. And Urban never spoke to Galileo again.

   A: I do not recall that this precept was intimated to me any
other way than by the voice of Lord Cardinal Bellarmino,

and I remember that the injunction was that I might not hold or defend; and there may have been also 'nor teach'. I do not remember that there was this phrase 'in any way', but there may have been; in fact I did not give any thought to it or keep it in mind because of my having, a few months later, that affidavit of the said Cardinal Bellarmino of the 26th of May which I have presented, in which is told the order to me not to hold or defend the said opinion. And the other two phrases now notified to me of the said precept, that is 'nor teach' and 'in any way', I have not kept in my memory, I think because they are not set forth in the said affidavit on which I relied, and which I have kept as a reminder.

Q: Whether, after the aforesaid injunction was issued to him, he obtained any permission to write the book he identified, which he later sent to the printer.

A: I did not seek permission to write the book, because I did not consider that in writing it I was acting contrary to, far less disobeying, the command not to hold, defend, or teach that opinion, but rather that I was refuting the opinion.

Q: Whether he obtained permission for printing the same book, by whom, and whether for himself or for someone else.

A: To obtain permission to print the above-mentioned book, although I was receiving profitable offers from France, Germany and Venice, I refused them and spontaneously came to Rome three years ago to place it into the hands of the chief censor, namely the Master of the Sacred Palace, giving him absolute authority to add, delete and change as he saw fit. After having it examined very diligently by his associate Father Visconti, the said Master of the Sacred Palace reviewed it again himself and licensed it; that is, having approved the book, he gave me permission but ordered to have the book printed in Rome. Since, in view of the approaching summer, I wanted to go home to avoid the danger of getting sick, having been away all of May and June,

The trial of Galileo

we agreed that I was to return here the autumn immediately following. While I was in Florence, the plague broke out and commerce was stopped; so, seeing that I could not come to Rome, by correspondence I requested of the same Master of the Sacred Palace permission for the book to be printed in Florence. He communicated to me that he would want to review my original manuscript, and that therefore I should send it to him. Despite having used every possible care and having contacted even the highest secretaries of the Grand Duke and the directors of the postal service, to try to send the said original safely, I received no assurance that this could be done, and it certainly would have been damaged, washed out or burned, such was the strictness at the borders. I related to the same Father Master this difficulty concerning the shipping of the book, and he ordered me to have the book again very scrupulously reviewed by a person acceptable to him; the person he was pleased to designate was Father Master Giacinto Stefani, a Dominican, professor of Sacred Scripture at the University of Florence, preacher for the Most Serene Highnesses, and consultant to the Holy Office. The book was handed over by me to the Father Inquisitor of Florence and by the Father Inquisitor to the said Father Giacinto Stefani; the latter returned it to the Father Inquisitor, who sent it to Signor Niccolò dell'Antella, reviewer of books to be printed for the Most Serene Highness of Florence; the printer, named Landini, received it from this Signor Niccolò and, having negotiated with the Father Inquisitor, printed it, observing strictly every order given by the Father Master of the Sacred Palace.

Q: Whether, when he sought permission from the Master of the Sacred Palace to print the said book, he revealed to the same Most Reverend Father Master the injunction previously given to him concerning the above-mentioned directive of the Holy Congregation.

A: I did not happen to discuss that command with the Master of the Sacred Palace when I asked for the imprimatur, for I did not think it necessary to say anything, because I had no doubts about it; for I have neither maintained nor defended in that book the opinion that the Earth moves and that the Sun is stationary but have rather demonstrated the opposite of the Copernican opinion and shown that the arguments of Copernicus are weak and inconclusive.

This last phrase of Galileo's testimony encapsulates the agony of his position. It would be easy to accuse him of equivocating. Surely by the end of that day's questioning he appreciated the danger he faced, and may have seen good reason to hedge in self-defence. Ambassador Niccolini had even warned him to be submissive and assume whatever attitude the inquisitors seemed to want of him. But Galileo did not lie under oath. He was a Catholic who had come to believe something Catholics were forbidden to believe. Rather than break with the Church, he had tried to hold – and at the same time not to hold – this problematic hypothesis, this image of the mobile Earth. His comment in the deposition recalls the duality he expressed in his 'Reply to Ingoli', when he described how Italian scientists had come to appreciate all the nuances of Copernicanism before rejecting the theory on religious grounds. That Galileo believed in his own innocence and sincerity is clear from letters he wrote before, during and long after the trial.

The prosecutors hearing Galileo's response, however, may well have gasped at it. Why had this case been referred to the Holy Office in the first place, if not because Urban's hired panel deemed the *Dialogue* an over-enthusiastic defence of Copernicus? The prosecutors could have questioned Galileo closely here on suspicion of deceit. But instead they said nothing. Perhaps they, too, understood the complexity of the situation. Or they took him at his word. Or both.

> With this the deposition ended, and he was assigned a certain room in the dormitory of the officials, located in the Palace of the Holy Office, in lieu of prison, with the injunction not to leave it without special permission, under penalty to be decided by the Holy Congregation; and he was ordered to sign below and was sworn to silence.
>
> I, Galileo Galilei, have testified as above.

[XXIII]

Vainglorious ambition, pure ignorance and inadvertence

WHILE GALILEO AWAITED THE outcome of this first hearing, confined to assigned rooms in the palace of the Inquisition, a second team of three theologians cross-examined the *Dialogue* itself. In less than a week, these consultors to the Holy Office, two of whom had served on the commission charged with reviewing the book the previous September, turned in statements of varying length and vehemence, all concurring that the book unabashedly backed Copernicus.

'It is beyond question that Galileo teaches the Earth's motion in writing,' concluded the Jesuit panelist, Melchior Inchofer. 'Indeed his whole book speaks for itself. Nor can one teach in any other way those of future generations and those who are absent except through writing . . . and he writes in Italian, certainly not to extend the hand to foreigners or other learned men, but rather to entice to that view common people in whom errors very easily take root.'

Not only did Inchofer submit the longest of the three condemnations of the *Dialogue*, but he also felt personally affronted by it. 'If Galileo had attacked some individual thinker for his inadequate arguments in favour of the stability of the Earth, we might still put a favourable construction on his text,' Inchofer said; 'but as he declares war on everybody and regards as mental dwarfs all who are not Pythagorean or Copernican, it is clear enough what he has in mind.'

Galileo had pleaded ignorance of the more harshly worded warning. Now the consultors claimed he had violated even the most liberal interpretation of the more lenient reproof – as in fact he had. Although the *Dialogue* displayed the imprimatur of the Sacred Palace, it reeked of heresy all the same, leaving the tribunal in a quandary the remainder of the month, trying to decide what must be done.

On 28 April, a memo from Father Commissary Vincenzo Maculano da Firenzuola reached the papal holiday retreat at Castel Gandolfo, where Urban had closeted himself with nephew Francesco Cardinal Barberini. Although the pope had instigated the trial of Galileo, Cardinal Barberini, as one of the ten inquisitor judges in it, expended every possible effort to protect his former mentor and fellow Lyncean Academician from his uncle Urban's anger. Perhaps Cardinal Barberini even suggested the very course of action that the commissary now reported having successfully accomplished – that is, he had convinced the Holy Congregation to let him deal extra-judicially with Galileo.

'And not to lose time,' the father commissary wrote to Cardinal Barberini, 'I went to reason with Galileo yesterday after luncheon, and after many exchanges between us I gained my point, by the grace of God; for I made him see that he was clearly wrong and that in his book he had gone too far.'

The commissary, a Dominican priest like Father Riccardi, but trained as a military engineer, well understood the virtues of the Copernican world-view. More than that, he person-

Image envisioning Galileo in prison

ally preferred to separate the construction of the universe from considerations of Holy Writ. But in that private tête-à-tête, he convinced Galileo to confess so as to let the affair end quietly with the least loss of face all around.

'The Tribunal will retain its reputation and be able to use benignity with the accused,' the commissary concluded his report to Cardinal Barberini. 'However things turn out, Galileo will recognise the grace accorded to him, and all the other satisfactory consequences that are wished for will follow.'

On Saturday, the last day of April, Galileo re-entered the commissary's chambers for a second formal hearing.

Over these intervening days of reflection, Galileo explained as he began the next set of his remarks recorded in the trial transcript, it had occurred to him to reread his *Dialogue*, which he had not looked at for the past three years. He meant to see whether, contrary to his own beliefs, something had perchance fallen from his pen to give offence.

'And, owing to my not having seen it for so long,' he explained, 'it presented itself to me like a new writing and by another author. I freely confess that in several places it seemed to me set forth in such a form that a reader ignorant of my real purpose might have had reason to suppose that the arguments brought on the false side, and which it was my intention to confute, were so expressed as to be calculated rather to compel conviction by their cogency than to be easy of solution.'

Here Galileo singled out his prized theories – the argument from the sunspots and the testimony of the tides – as having been presented far too powerfully, when in fact they provided no proof. He supposed he had succumbed to 'the natural complacency which every man feels with regard to his own subtleties and for showing himself to be more skilful than the generality of men in devising, even in favour of false propositions, ingenious and plausible arguments.

'My error, then, has been – and I confess it – one of vainglorious ambition and of pure ignorance and inadvertence.'

Dismissed, he left the room at that point, the record shows, but put his head in the door moments later asking permission to show his good faith, as he now felt ready to do, by softening his stance on Copernicus: 'And there is a most favourable opportunity for this, seeing that in the work already published the interlocutors agree to meet again after a certain time to discuss several distinct problems of Nature not connected with the matter already treated. As this affords me an opportunity of adding one or two other "Days", I promise to resume the arguments already brought in favour of the said opinion, which is false and has been condemned, and to refute them in such an effectual way as by the blessing of God may be supplied to me. I pray, therefore, this Holy Tribunal to aid me in this good resolution and to enable me to put it into effect.'

By this suggestion, Galileo apparently hoped to save his *Dialogue* from being banned.

After hearing this heartfelt plea, the commissary returned Galileo to the Tuscan embassy, out of consideration for the arthritic pains that now tormented the old man more than usual.

'It is a fearful thing to have to do with the Inquisition,' Ambassador Niccolini observed after welcoming Galileo once again at Villa Medici. 'The poor man has come back more dead than alive.'

The Inquisition had not yet decided Galileo's fate, by any

means, and still retained the power to have him tortured or imprisoned. Nevertheless, the change of venue, which Galileo immediately communicated to his friends and family, bathed them all in its sweet reprieve.

Most Illustrious and Beloved Lord Father

THE DELIGHT DELIVERED TO me by your latest loving letter was so great, and the change it wrought in me so extensive, that, taking the impact of the emotion together with my being compelled many times to read and reread the same letter over and over to these nuns, until everyone could rejoice in the news of your triumphant successes, I was seized by a terrible headache that lasted from the fourteenth hour of the morning on into the night, something truly outside my usual experience.

I wanted to tell you this detail, not to reproach you for my small suffering, but to enable you to understand all the more how heavily your affairs weigh on my heart and fill me with concern, by showing you what effects they produce in me; effects which, although, generally speaking, filial devotion can and should produce in all progeny, yet in me, I will dare to boast that they possess greater force, as does the power that places me far ahead of most other daughters in the love and reverence I bear my dearest Father, when I see clearly that he, for his part, surpasses the majority of fathers in loving me as his daughter: and that is all I have to say.

I offer endless thanks to blessed God for all the favours and graces that you have been granted up till now, Sire, and hope to receive in the future, since most of them issue from that merciful hand, as you most justly recognise. And even though you attribute the great share of these blessings to the merit of my prayers, this truly is little or nothing; what matters most is the sentiment with which I speak of you to His Divine Majesty, Who, respecting that love, rewarding you so beneficently, answers my prayers, and renders us ever more greatly obligated to Him, while we are also deeply indebted to all those people who have given you their goodwill and aid, and especially to those most pre-eminent nobles who are your hosts. And I did want to write to Her Most Excellent Ladyship the Ambassadress, but I stay my hand lest I vex her with my constant repetition of the same statements, these being expressions of thanks and confessions of my infinite indebtedness. You take my place, Sire, and pay respects to her in my name. And truly, dearest lord Father, the blessing that you have enjoyed from the favours and the protection of these dignitaries is so great that it suffices to assuage, or even annul all the aggravations you have endured.

Here is a copy I made you of a most excellent prescription against the plague that has fallen into my hands, not because I believe there is any suspicion of the malady where you are, but because this remedy also works well for all manner of ills. As to the ingredients, I am in such short supply that I must beg them for myself, on which account I cannot fill the prescription for anyone else; but you must try to procure those ingredients that perchance you may lack, Sire, from the heavenly foundry, from the depths of the compassion of the Lord God, with Whom I leave you. Closing with regards to you from everyone here, and in particular from Suor Arcangela and Suor Luisa, who for now, as far as her health is concerned, is getting along passing well.

FROM SAN MATTEO, THE 7TH DAY OF MAY 1633.
Most affectionate daughter,

S. M. Celeste

The hard-to-fill prescription for plague preventive that Suor Maria Celeste enclosed on a separate slip of paper has not survived with her letter. Perhaps Galileo lost the recipe as a result of keeping it with him for luck or inspiration. No doubt it called for heroic measures of faith, strength, virtue and acceptance of Divine Will – and none would deny the need for all of these in the days ahead.

On 10 May, Galileo returned for his third deposition in the commissary's chambers, to tender his formal written defence.

'In an earlier investigation', Galileo's statement began, 'I was asked whether I had informed the Most Reverend Father Master of the Sacred Palace about the private injunction issued to me sixteen years ago by order of the Holy Office – 'not to hold, defend, or teach in any way whatever' the opinion of the Earth's motion and the Sun's stability – and I answered No. Since I was not asked the reason why I did not inform him, I did not have the opportunity to say anything else. Now it seems to me necessary to mention it, in order to prove the absolute purity of my mind, always averse to using simulation and deceit in any of my actions.'

He then reviewed the events of 1616 that led to his asking Cardinal Bellarmino for the affidavit that he attached herewith as evidence. 'In it one clearly sees that I was only told not to hold or defend Copernicus's doctrine of the Earth's motion and the Sun's stability; but one cannot see any trace that, besides the general pronouncement applicable to all, I was given any other special order.'

Since Cardinal Bellarmino's private communication precisely matched the terms of the public Edict of 1616, Galileo argued, and neither document contained the words *nor teach* or *in any way whatever*, these phrases now struck him as 'very new and unheard'. He should not be mistrusted, he said, for having forgotten, 'in the course of fourteen or sixteen years', whether such terms had ever been uttered in his presence. And really, he had seen no need to notify the master of the Sacred Palace of his private injunction from Cardinal Bellarmino, when it said nothing more specific than the published public one.

'Given that my book was not subject to more stringent censures than those required by the decree of the Index,' Galileo continued, 'I followed the surest and most effective way to protect it and purge it of any trace of blemish. It seems to me that this is very obvious, since I handed it over to the supreme Inquisitor at a time when many books on the same subjects were being prohibited solely on account of the above-mentioned decree.'

Therefore, he hoped the 'Most Eminent and Most Prudent Lord Judges' would concede that he had neither willfully nor knowingly disobeyed any orders given him. Indeed, 'those flaws that can be seen scattered in my book were not introduced through the cunning of an insincere intention, but rather through the vain ambition and satisfaction of appearing clever above and beyond the average among popular writers'.

After he logically ordered all the details of his affair, explained his thinking, defended the purity of his intention and declared himself ready to make amends, Galileo pleaded for mercy. 'Lastly, it remains for me to pray you to take into consideration my pitiable state of bodily indisposition, to which, at the age of seventy years, I have been reduced by ten months of constant mental anxiety and the fatigue of a long and toilsome journey at the most inclement season – together with the loss of the greater part of the years of which, from my previous condition of health, I had the prospect.' He hoped his judges would deem his decrepitude and disabilities adequate punishment for his mistakes.

'Equally,' he concluded, 'I want them to consider my honour and reputation against the slanders of those who hate me.'

Over the next several weeks, while the tribunal prepared its final report to the pope, Galileo resumed his now familiar occupation of waiting in suspense at the Tuscan embassy. Unbeknownst to him, the grand duke had decided to stop paying Galileo's bills there, expecting him to cover his own expenses henceforward. Ferdinando offered no explanation for this uncharacteristically ungenerous action, but he may have been moved by Urban's warnings in the Galileo affair, by the April incarceration of another Florentine citizen – Mariano Alidosi – in the prisons of the Holy

Office, and by the fiscal strain of paying for the ravages of the plague. Whatever the reason, Ambassador Niccolini took the news badly.

'In regard to what Your Most Illustrious Lordship tells me,' he wrote to the secretary of state on 15 May, 'namely that His Highness does not intend to pay for Signor Galilei's expenses here beyond the first month, I can reply that I am not about to discuss this matter with him while he is my guest; I would rather assume the burden myself.'

The ambassador made this pledge in the full knowledge that his 'guest' might remain under house arrest in his villa for as long as six months – until the trial process lumbered to its untimely resolution.

Urban had returned to Rome meanwhile from Castel Gandolfo and reinserted himself into the disposition of Galileo's case. He could see immediately how the inquisitors of the Holy Office fell into pro- and anti-Galileo factions: some of them had actually attempted to read the *Dialogue* and came away from it feeling enlightened; others bristled at the dishonesty they perceived in Galileo's pathetic defence. All of them, however, agreed that Galileo had, at the very least, disobeyed direct orders.

Even if Urban's former love for Galileo had remained untarnished, instead of being tainted by betrayal, he could not now deny the obvious. The accused had defended a condemned doctrine. Nor could Urban risk any overt leniency towards Galileo, considering how the Thirty Years' War had raised doubts about his own guardianship of the Catholic faith. And no matter how much Urban admired the achievements of Galileo's lifetime, he had never shared his outlook on the ultimate goal of scientific discovery. Whereas Galileo believed that Nature followed a Divine order, which revealed its hidden pattern to the persistent investigator, Urban refused to limit his omnipotent God to logical consistency. Every effect in Nature, as the handiwork of God, could claim its own fantastic foundation, and each of these would necessarily exceed the limits of human imagining – even of a mind as gifted as Galileo's.

[XXIV]

Faith vested in the miraculous Madonna of Impruneta

ALL THE WHILE GALILEO sojourned in Rome, the plague regathered its strength in Florence. Suor Maria Celeste heard regularly from Signor Rondinelli of new developments and new cases reported to the local Magistracy of Public Health. Thus, as much as she pined to have her father back at Il Gioiello, she counselled him, out of concern for his welfare, to enjoy the hospitality of his Roman hosts as long as possible. She even suggested he celebrate his eventual release from the Holy Office by making another pilgrimage to the Casa Santa in Loreto – anything to delay his homecoming until the pestilence deserted Tuscany altogether.

The walls of San Matteo continued to keep the plague at bay. Inside them, other sicknesses assailed the sisters, and all manner of non-health-related problems as well, such as the débâcle of Suor Arcangela's term as provider. In this role, which fell to each of the nuns on a rotating basis, Suor Arcangela assumed charge of all outside food purchases for the convent for one year. As little as the nuns ate, the cost of twelve months' provisions for thirty women could easily reach one hundred *scudi*, most of which the provider had to pay out of her own pocket, even if that pocket were empty. Some previous providers had resorted to desperate schemes to produce the requisite sums, and a few had circumvented Suor Maria Celeste to appeal directly to Galileo for emergency assistance. Now his own daughter found herself in this plight as her term of office neared its end. But just as Suor Arcangela had prevailed on others to help her shoulder the physical demands of the job, owing to her frequent spells and infirmities, she deferred the worrisome financial burden as well.

'I have another kindness to ask of you,' Suor Maria Celeste was thus forced to address Galileo in April while he faced foes far from home, 'not for myself, but for Suor Arcangela, who, by the grace of God, three weeks from today, which will be the last day of this month, must leave the office of Provider, in which capacity up till now she has spent one hundred *scudi* and fallen into debt; and being obliged to leave 25 *scudi* in reserve for the new Provider, not having anywhere else to turn, I beg your leave, with your permission, Sire, to help her out with the money of yours that I hold, so that this ship can bring itself safely into port, whereas truly, without your help, it would not complete so much as half its voyage.'

Even after Galileo approved the release of those funds, however, Suor Maria Celeste had to approach him for more. 'For Suor Arcangela up till now I have paid close to 40, part of which I had received as a loan from Suor Luisa, and part from our allowance, of which there remains a balance of 16 *scudi* to draw upon for all of May. Suor Oretta spent 50 *scudi*: now we are sorely pressed and I do not know where else to turn, and since the Lord God keeps you in this life for our support, I take advantage of this blessing and seize upon it to beseech you, Sire, that with God's love you free me from the worry that harasses me, by lending me whatever amount of money you can until next year comes around, at which time we will recover our losses by collecting from those who must pay the expenses then, and thereby repay you.'

On the back of this letter Galileo wrote himself a reminder – of something he might not easily forget – as though to aid a memory already taxed by age and stress. The note said, 'Suor Maria Celeste needs money immediately.' And, as usual, he saw to that need.

As soon as she had settled her sister's debts with their father's help, Suor Maria Celeste learned that Suor Arcangela's next proposed position would put her in charge of the convent's wine cellar. 'This current year was to bring Suor Arcangela's turn as Cellarer, an office that gave me much to ponder,' Suor Maria Celeste told Galileo. She feared Suor Arcangela would abase the position as she had that of provider, through inattention, and perhaps abuse it, too, by usurping her authority as cellarer to overindulge in drink. 'Indeed I secured the Mother Abbess's pardon that it not be given to her by pleading various excuses; and instead she was made Draper, obliging her to bleach and keep count of the tablecloths and towels in the Convent.'

When she heard, probably from Galileo himself, how freely fine vintages flowed at the Niccolinis' table, Suor Maria Celeste admonished her father not to confuse himself with wine – the beverage he had long fondly described as 'light held together by moisture'. She feared the large quantities he'd grown accustomed to imbibing might aggravate his worsening pains.

'The evil contagion still persists,' she wrote on 14 May in response to Galileo's hint of how he might soon be headed homeward, 'but they say only a few people die of it and the hope is that it must come to an end when the Madonna of Impruneta is carried in procession to Florence for this purpose.'

The Madonna of Impruneta, an icon brought to Tuscany in the

The Madonna of Impruneta

first century, had ministered to the victims of sundry disasters. Since the Black Death in 1348, not a flood nor a drought nor a famine, battle or epidemic had erupted in Florence that could not be mitigated by carrying the miraculous Madonna of Impruneta from her small church in a neighbouring village down into the great cathedrals of the city. In May of 1633, Grand Duke Ferdinando called for the sacred image to retrace that long path once more, to rout the resurgent plague. The Magistracy of Public Health, however, attentive to the danger of congregation in time of contagion, issued an ordinance on 20 May to limit the number of faithful – especially women and children – who could join the procession.

'As for your returning here under these prevailing conditions,' Suor Maria Celeste worried in her next letter, 'I can guarantee you neither resolution nor assurance on account of the contagious pestilence, whose end is so urgently desired that all the faith of the city of Florence is now vested in the Most Holy Madonna, and to this effect this morning with great solemnity her miraculous Image was carried from Impruneta to Florence, where it is expected to stay for 3 days, and we cherish the hope that during its return journey we will enjoy the privilege of seeing her.'

From 20 to 23 May, the Madonna of Impruneta passed through the streets of Florence laid waste by the plague, and spent her nights as the guest of three churches that had vied for that honour. The holy image depicted Mary seated on a throne, wearing a red dress and jewelled crown, holding the baby Jesus in the folds of her blue lap robe. About her neck, real necklaces – one of pearls, the other of precious stones – embossed the flat surface of the icon, the whole of which stood inside a red frame rounded at the top like an archway. Eight men carried the poles holding the ornate canopy above the Madonna, and at least another twelve shared the weight and the glory of bearing the table on which her venerated form rested.

When leaving Florence, Suor Maria Celeste recounted, 'the image of the Most Holy Madonna of Impruneta came into our Church; a grace truly worthy of note, because she was passing from the Piano [dei Giullari], so that she had to come this way, going back along the whole length of that road you know so well, Sire, and weighing in excess of 700 *libbre* [about one-quarter ton] with the tabernacle and adornments; its size rendering it unable to fit through our gate, it became necessary to break the wall of the courtyard, and raise the doorway of the Church, which we accomplished with great readiness for such an occasion.'*

In the weeks following the Madonna's visit, the death toll from the plague alternately dipped and climbed, so that the full extent of the miracle she had wrought did not become clear until 17 September 1633, when the authorities declared Florence officially free of contamination. In early June, however, as the day of Galileo's judgment approached, Suor Maria Celeste still heard word of seven or eight plague fatalities every day. And Signor Rondinelli warned her that the open country around Rome would be closed

---

* Even today, the Madonna is kept hidden during ordinary times. Visitors to the Impruneta church, which was rebuilt following bomb hits during World War II, must content themselves with simply being near the icon, as it reposes inside a marble shrine, behind a blue gilt-embroidered curtain.

to travellers for the summer months as a further precaution to preserve the safety of the Holy City. If Galileo were not allowed to leave the embassy in the very near future, he would be stuck there at least until autumn.

In mid-June she again cautioned her father against hasty departure, urging him to stay put, away from 'these perils, which in spite of everything continue and may even be multiplying; and in consequence an order has come to our Monastery, and to others as well, from the Commissioners of Health, stating that for a period of 40 days we must, two nuns at a time, pray continuously day and night beseeching His Divine Majesty for freedom from this scourge. We received alms of 25 *scudi* from the commissioners for our prayers, and today marks the fourth day since our vigil began.'

All the worry Suor Maria Celeste expended over the plague danger facing her father was unoccasioned, however, since the Inquisition had no intention of releasing Galileo any time soon.

On 16 June, Pope Urban VIII presided over a meeting of the cardinal inquisitors. Urban had absorbed the official report summarising the Galileo affair from the first accusations against the philosopher in 1615, through the publication of his book, up to his recent defence and plea for mercy. Now His Holiness demanded that Galileo be interrogated 'on intent' – to determine, technically by torture if necessary, his true purpose in writing the *Dialogue*. The book itself could not escape censure in any case, the pontiff averred, and would assuredly be prohibited. As for Galileo, he would have to serve a prison term and perform penance. His public humiliation would warn all Christendom of the folly of disobeying orders and gainsaying Holy Scripture dictated by the mouth of God.

'To give you news of everything about the house', Suor Maria Celeste wrote on 18 June, unaware of the awful turn of events in Rome,

> I will start from the dovecote, where since Lent the pigeons
> have been brooding; the first pair to be hatched were

devoured one night by some animal, and the pigeon who had been setting them was found draped over a rafter half eaten, and completely eviscerated, on which account La Piera assumed the culprit to be some bird of prey; and the other frightened pigeons would not go back there, but, as La Piera kept on feeding them they have since recovered themselves, and now two more are brooding.

The orange trees bore few flowers, which La Piera pressed, and she tells me she has drawn a whole pitcherful of orange water. The capers, when the time comes, will be sufficient to suit you, Sire. The lettuce that was sown according to your instructions never came up, and in its place La Piera planted beans that she claims are quite beautiful, and coming lastly to the chickpeas, it seems the hare will win the largest share, he having already begun to make off with them.

The broad beans are set out to dry, and their stalks fed for breakfast to the little mule, who has become so haughty that she refuses to carry anyone, and has several times thrown poor Geppo so as to make him turn somersaults, but gently, since he was not hurt. Sestilia's brother Ascanio once asked to ride her out, though when he approached the gate to Prato he decided to turn back, never having gained the upper hand over the obstinate creature to make her proceed, as she perhaps disdains to be ridden by others, finding herself without her true master.

On the morning of 21 June, ushered into the chambers of the commissary general for the fourth and final time, Galileo endured his examination on intent by Father Maculano.

Q: Whether he had anything to say.

A: I have nothing to say.

Q: Whether he holds or has held, and for how long, that the Sun is the centre of the world and the Earth is not the centre of the world but moves also with diurnal motion.

A: A long time ago, that is, before the decision of the Holy Congregation of the Index, and before I was issued that injunction, I was undecided and regarded the two opinions, those of Ptolemy and Copernicus, as disputable, because either the one or the other could be true in Nature. But after the said decision, assured by the prudence of the authorities, all my uncertainty stopped, and I held, as I still hold, as most true and indisputable, Ptolemy's opinion, namely the stability of the Earth and the motion of the Sun.

Having been told that he is presumed to have held the said opinion after that time, from the manner and procedure in which the said opinion is discussed and defended in the book he published, indeed from the very fact that he wrote and published the said book, he was asked therefore to freely tell the truth whether he holds or has held that opinion.

A: In regard to my writing of the published *Dialogue*, I did not do so because I held the Copernican doctrine to be true. Instead, deeming only to confer a common benefit, I set forth the physical and astronomical reasons that can be advanced for each side; I tried to show that neither set of arguments has the force of conclusive demonstration in favour of the one opinion or the other, and that therefore to proceed with certainty one had to resort to the decisions of higher teaching, as one can see in many passages in the *Dialogue*. So for my part I conclude that I do not hold and, after the determination of the authorities, I have not held the condemned opinion.

Having been told that from the book itself and the reasons advanced for the affirmative side, namely that the Earth moves and the Sun is motionless, he is presumed, as it was stated, to hold Copernicus's opinion, or at least to have held it at the time, therefore he was told that unless he decided to proffer the truth, one would have recourse to the remedies of the law and to appropriate steps against him.

A: I do not hold this opinion of Copernicus, and I have not held it after being ordered by injunction to abandon it. For the rest, I am here in your hands; do with me what you please.

And he was told to tell the truth, otherwise one would have recourse to torture.

A: I am here to obey, but I have not held this opinion after the determination was made, as I said.

[XXV]

Judgment passed on your book and your person

DESPITE THE HOPES OF Galileo and his supporters that his affair would end quietly in a private admonition – with his *Dialogue* merely 'suspended until corrected', as Copernicus's book had been – the sentence pronounced on Wednesday 22 June publicly convicted him of heinous crimes.

The cardinal inquisitors and their witnesses gathered that morning in the Dominican convent adjoining the Church of Santa Maria Sopra Minerva, in the centre of the city, where they typically held their weekly meetings. Up a spiral staircase, into a room with a frescoed ceiling came Galileo, led before them to hear the results of their deliberations.

We say, pronounce, sentence, and declare that you, Galileo, by reason of the matters which have been detailed in the trial and which you have confessed already, have rendered yourself in the judgment of this Holy Office vehemently suspected of heresy, namely of having held and believed the doctrine which is false and contrary to the Sacred and Divine Scriptures, that the Sun is the centre of the world and does not move from east to west and that the Earth moves and is not the centre of the world; and that one may hold and defend as probable an opinion after it has been declared and defined contrary to Holy Scripture. Consequently, you have incurred all the censures and penalties enjoined and promulgated by the sacred Canons and all particular and general laws against such delinquents. We are willing to absolve you from them provided that first, with a sincere heart and unfeigned faith, in our presence you abjure, curse and detest the said errors and heresies, and every other error and heresy contrary to the Catholic and Apostolic Church in the manner and form we will prescribe to you.

Furthermore, so that this grievous and pernicious error and transgression of yours may not go altogether unpunished, and so that you will be more cautious in future, and an example for others to abstain from delinquencies of this sort, we order that the book *Dialogue of Galileo Galilei* be prohibited by public edict.

We condemn you to formal imprisonment in this Holy Office at our pleasure. As a salutary penance we impose on you to recite the seven penitential psalms once a week for the next three years. And we reserve to ourselves the power of moderating, commuting, or taking off, the whole or part of the said penalties and penances. This we say, pronounce, sentence, declare, order and reserve by this or any other better manner or form that we reasonably can or shall think of. So we the undersigned Cardinals pronounce.

Even though the opinion of Copernicus had been rescued from the shame of heresy in 1616, Galileo, for his exposition of Copernicus, now stood 'vehemently suspected of heresy' himself.

Only seven of the ten inquisitors affixed their signatures to the sentence. Francesco Cardinal Barberini, the strongest advocate for clemency among them, pointedly stayed away from the session and declined to sign. Also absent was Gaspare Cardinal Borgia, who perhaps used this occasion to reproach Pope Urban further for his pro-French behaviour in the Thirty Years' War – or to thank Galileo for the suggestions he once offered the Spanish government on solving the longitude problem by observing the moons of Jupiter. Laudivio Cardinal Zacchia, one of the first cardinals to whom Grand Duke Ferdinando wrote in defence of Galileo, also withheld his signature, for reasons equally unknown. Perhaps he was ill that day and could not attend.

The Holy Tribunal presented Galileo its draft text of an abjuration for him to speak aloud. But in reading it first silently to himself, he discovered two clauses so abhorrent that he could not be convinced, even under the circumstances, to concede them: one suggested he had lapsed in his behaviour as a good Catholic, the other that he had acted deceitfully in obtaining the imprimatur for the *Dialogue*. He had done nothing of the kind, he said, and the officials granted his request to strike these references from the script.

Dressed in the white robe of the penitent, the accused then knelt and abjured as ordered:

> I, Galileo, son of the late Vincenzio Galilei, Florentine, aged 70 years, arraigned personally before this tribunal, and kneeling before You, Most Eminent and Reverend Lord Cardinals, Inquisitors-General against heretical depravity throughout the Christian commonwealth, having before my eyes and touching with my hands the Holy Gospels, swear that I have always believed, I believe now, and with God's help I will in future believe all that is held, preached, and taught by the Holy Catholic and Apostolic Church. But whereas – after

having been admonished by this Holy Office entirely to abandon the false opinion that the Sun is the centre of the world and immovable, and that the Earth is not the centre of the same and that it moves, and that I must not hold, defend, nor teach in any manner whatever, either orally or in writing, the said false doctrine, and after it had been notified to me that the said doctrine was contrary to Holy Writ – I wrote and caused to be printed a book in which I treat of the already condemned doctrine, and adduce arguments of much efficacy in its favour, without arriving at any solution: I have been judged vehemently suspected of heresy, that is, of having held and believed that the Sun is the centre of the world and immovable, and that the Earth is not the centre and moves.

Therefore, wishing to remove from the minds of your Eminences and of all faithful Christians this vehement suspicion justly conceived against me, I abjure with a sincere heart and unfeigned faith, I curse and detest the said errors and heresies, and generally all and every error and sect contrary to the Holy Catholic Church. And I swear that for the future I will never again say nor assert in speaking or writing such things as may bring upon me similar suspicion; and if I know any heretic, or person suspected of heresy, I will denounce him to this Holy Office, or to the Inquisitor or Ordinary of the place where I may be. I also swear and promise to adopt and observe entirely all the penances which have been or may be imposed on me by this Holy Office. And if I contravene any of these said promises, protests or oaths (which God forbid!), I submit myself to all the pains and penalties imposed and promulgated by the Sacred Canons and other Decrees, general and particular, against such offenders. So help me God and these His Holy Gospels, which I touch with my own hands.

I, the said Galileo Galilei, have abjured, sworn, promised, and bound myself as above; and in witness of the truth, with my own hand have subscribed the present document of my abjuration, and have recited it word by word in Rome, at the Convent of the Minerva, this 22nd day of June 1633.

I, Galileo Galilei, have abjured as above, with my own hand.

It is often said that as Galileo rose from his knees he muttered under his breath, '*Eppur si muove*' (But still it moves). Or he shouted out these words, looking towards the sky and stamping his foot. Either way, for Galileo to voice such undaunted conviction in this hostile encounter would have been beyond foolhardy, not to mention that the comment suggests a defiant feistiness beyond his means to muster then and there. He may have said it weeks or months later, in front of other witnesses, but not on that day. He sustained his condemnation in the Convent of the Minerva as a breach of the promises made him in exchange for his cooperation. For he believed in his own innocence; he had admitted committing a 'crime' only because his confession had been part of a deal.

Within days, Cardinal Barberini successfully softened Galileo's sentence by changing the place of his imprisonment from the

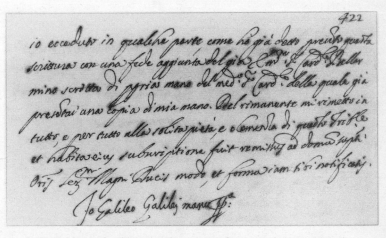

Final lines of Galileo's handwritten confession to the Inquisition

dungeons of the Holy Office to the Tuscan embassy in Rome. Then Ambassador Niccolini entreated Pope Urban to pardon Galileo and send him home to Florence. Galileo, he explained to bolster his plea, had agreed to take in his widowed sister-in-law, who was even now preparing to depart from Germany along with her eight children, and had nowhere else to turn.

Urban rejected the idea of the pardon, but he consented to let Galileo leave Rome at last. With Cardinal Barberini's intervention, Galileo was consigned for the first five months of his prison term to the custody of the archbishop of Siena, who had already offered to send his personal litter to ensure safe, speedy conveyance to his palace.

The *Dialogue* duly appeared on the next published Index of Prohibited Books, in 1664, where it would remain for nearly two hundred years.

Most Illustrious and Beloved Lord Father

JUST AS SUDDENLY AND unexpectedly as word of your new torment reached me, Sire, so intensely did it pierce my soul with pain to hear the judgment that has finally been passed, denouncing your person as harshly as your book. I learned all this by importuning Signor Geri, because, not having any letters from you this week, I could not calm myself, as though I already knew all that had happened.

My dearest lord father, now is the time to avail yourself more than ever of that prudence which the Lord God has granted you, bearing these blows with that strength of spirit which your religion, your profession, and your age require. And since you, by virtue of your vast experience, can lay claim to full cognisance of the fallacy and instability of everything in this miserable world, you must not make too much of these storms, but rather take hope that they will soon subside and transform themselves from troubles into as many satisfactions.

In saying all that I am speaking what my own desires dictate, and also what seems a promise of leniency demonstrated towards you, Sire, by His Holiness, who has destined for your prison a place so delightful, whereby it appears we may anticipate another commutation of your sentence conforming even more closely with all your and our wishes; may it please God to see things turn out that way, if it be for the best. Meanwhile I pray you not to leave me without the consolation of your letters, giving me reports of your condition, physically and especially spiritually: though I conclude my writing here, I never cease to accompany you with my thoughts and prayers, calling on His Divine Majesty to grant you true peace and consolation.

FROM SAN MATTEO THE 2ND DAY OF JULY 1633.
Most affectionate daughter,

S. M. Celeste

Galileo's humiliation spread outwards from Rome as fast as messengers could carry the news. At the pope's command, and with fanfare, the text of Galileo's mortifying sentence was posted and read aloud by inquisitors from Padua to Bologna, from Milan to Mantua, from Florence to Naples to Venice and on to France, Flanders and Switzerland, alerting the professors of philosophy and mathematics at each locale of the outcome of Galileo's affair. In Florence in particular, instructions called for the condemnation to occasion a plenary session of the Florentine Inquisition, to which as many mathematicians as possible should be invited.

In the summer of 1633, spurred by these events, a lively black market trade sprang up around the banned *Dialogue*. The price of the book, which had originally sold for half a *scudo*, rose to four and then to six *scudi* as priests and professors across the country purchased copies to keep the inquisitors from cornering the market. A Paduan friend of Galileo's wrote to tell him how the well-known university philosopher Fortunio Liceti had actually surrendered his copy to the authorities, as though this were a most singular aberrant act, and no one else who owned the *Dialogue* could bear to part with it. As the book grew in esteem among Galileo's cohorts, it also gained new converts.

Some time later that July or August, a messenger smuggled a copy of the *Dialogue* across the Alps, with the help of clandestine agents, to Strasbourg, where the Austrian historian Mathias Berneggar began to prepare a Latin translation that would be ready for general distribution throughout Europe by 1635. In 1661, an English version of the *Dialogue* appeared, translated by Thomas Salusbury. The prohibition of the book by the Index endured for a long time, but not with a long arm outside Italy.

In 1744, publishers in Padua gained permission to include the *Dialogue* in a posthumous collection of Galileo's works, by inserting appropriate qualifying remarks and disclaimers in the text. But this concession did not lead to any rapid relaxation of the ruling against the *Dialogue*, which remained officially prohibited. On 16 April 1757, when the Congregation of the Index withdrew its general objections against books teaching the Copernican doctrine, it specifically cited the *Dialogue* as a still-forbidden title. The *Dialogue* in fact stayed banned for another sixty-five years, until 1822, when the Congregation of the Holy Office decided to allow publication of books on modern astronomy expounding the movement of the Earth. No new Index, however, was issued at the time to reflect this change in attitude. Thus the 1835 edition became the first Index in almost two centuries to drop the listing of the *Dialogue of Galileo Galilei*.

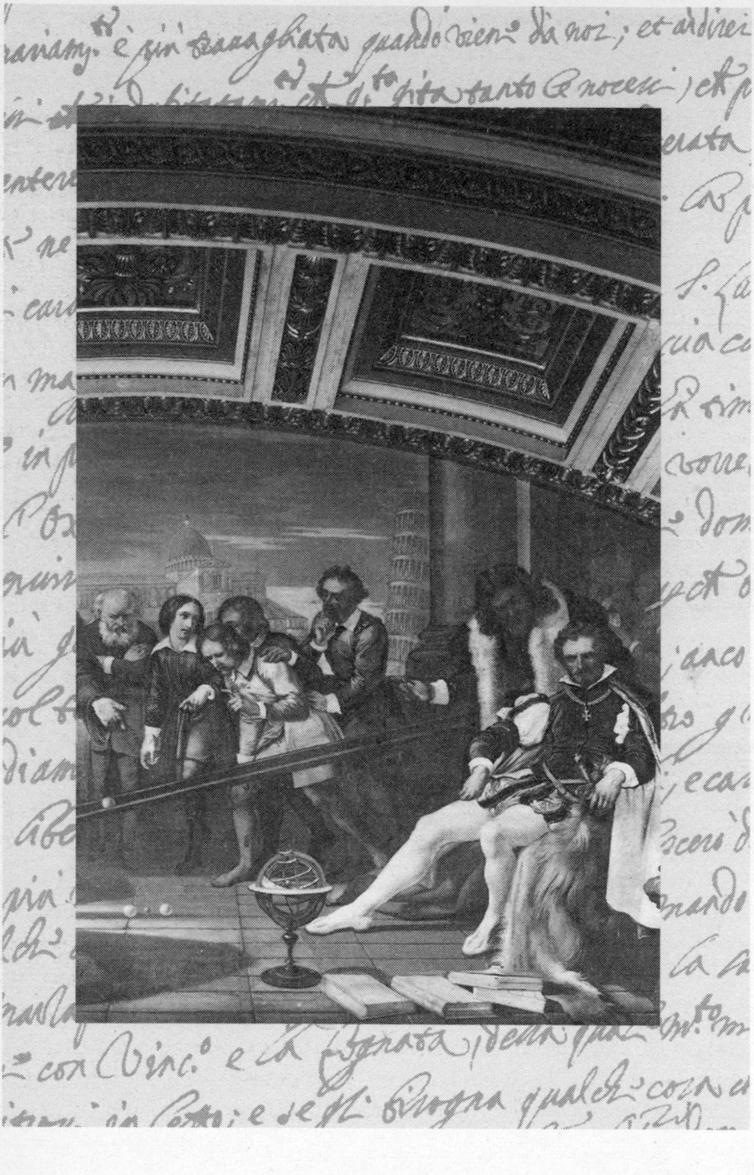

PART FIVE

At Siena

[XXVI]

Not knowing how to refuse him the keys

IN FRAMING GALILEO'S TRIAL as a simplistic case of science versus religion, anti-Catholic critics have claimed that the Church opposed a scientific theory on biblical grounds, and that the outcome mocked the infallibility of the pope. Technically, however, the anti-Copernican Edict of 1616 was issued by the Congregation of the Index, not by the Church. Similarly, in 1633, Galileo was tried and sentenced by the Holy Office of the Inquisition, not by the Church. And even though Pope Paul V approved the Edict of

1616, just as Pope Urban VIII condoned Galileo's conviction, neither pontiff invoked papal infallibility in either situation. The freedom from error that belongs to the pope as his special privilege applies only when he speaks as shepherd of the Church to issue formal proclamations on matters of faith and morals. What's more, the right of infallibility was never formally defined in Galileo's time, but issued two centuries later from Vatican Council I, held in 1869–70.

Although Urban personally believed in his own power enough to boast that the sentence of one living pope – namely Urban himself – outweighed all the decrees of one hundred dead ones, he refrained from claiming infallibility in the Galileo case.

French philosopher René Descartes, who followed Galileo's trial from his home in Holland, understood these distinctions. Thus Descartes could comment hopefully to a colleague: 'As I do not see that this censure has been confirmed either by a Council or by the Pope, but proceeds solely from a committee of cardinals, it may still happen to the Copernican theory as it did to that of the antipodes [the eighth-century censured notion of a sub-Earth Earth with human inhabitants] which was once condemned in the same way.' Nevertheless Descartes, a product of his Jesuit education and a devout Catholic throughout his life, withheld from publishing the book he himself had just completed, *Le Monde*, which espoused Copernicus's view of the universe.

Numerous churchmen, however – including highly placed clerics such as Ascanio Piccolomini, the archbishop of Siena – had endorsed Galileo's *Dialogue* from their initial reading of it and continued to consider the author their friend. 'It seems exceeding strange to me', Archbishop Piccolomini had written to Galileo when the outcry over the *Dialogue* first arose, 'that such a recent and precise approbation should be opposed by the passions of some people who might find fault only in what they conceive of the book, for the work itself ought to appease the most timid conscience. On the other hand, I will say you deserve this and worse, for you have been disarming by steps those who have control of the sciences, and they have nothing left but to run back to holy ground.'

Archbishop Piccolomini, capping a long line of scholars from a distinguished family that produced two popes, had himself studied mathematics and been Galileo's admirer for many years. Now, in the aftermath of the trial, Piccolomini assumed custody of Galileo, who left the Tuscan embassy in Rome on 6 July and arrived at the archiepiscopal palace, immediately adjacent to the magnificent domed cathedral of Siena, three days later on 9 July.

Ascanio Piccolomini, Archbishop of Siena

If Galileo looked 'more dead than alive', as Ambassador Niccolini had described him when he returned from his April questioning by the Holy Office, how must he have appeared at Siena in July, after these many more weeks and traumas? The injustice of the sentence tormented him so that he did not sleep for several nights but could be heard crying out, babbling and rambling in distraction. The archbishop feared for his safety to the point where he considered binding Galileo's arms to keep him from accidentally injuring himself in bed. Piccolomini determined to restore the man's broken spirit and return his thoughts to scientific pursuits. The fact that Galileo was able to rise from the ashes of his condemnation by the Inquisition to complete another book project (and for this last work to prove his greatest original contribution) is due in large measure to Piccolomini's solicitous kindness. A French visitor to Siena in 1633, the poet Saint-Amant, reported finding the archbishop and Galileo together among the rich tapestries and furnishings that filled the guest apartment at the palace, engaging each other in discussion of a mechanical theory, which lay partially written on pages spread all around them.

Most Illustrious and Beloved Lord Father

THAT THE LETTER YOU wrote me from Siena (where you say you find yourself in good health) brought me the greatest pleasure, and the same to Suor Arcangela, is needless for me to weary myself in convincing you, Sire, since you will well know how to fathom what I could not begin to express; but I should love to describe to you the show of jubilation and merriment that these mothers and sisters made upon learning of your happy return, for it was truly extraordinary; since the Mother Abbess, with many others, hearing the news, ran to me with open arms, and crying with tenderness and happiness; truly I am bound as a slave to all of them, for having understood from this display how much love they feel for you, Sire, and for us.

Hearing furthermore that you are staying in the home of a host as kind and courteous as Monsignor Archbishop multiplies the pleasure and satisfaction, despite the potential prejudicial effect this may have on our own interests, because it could well prove to be the case that his extremely enjoyable conversation may engage and detain you much longer than we would like. However, since here for now the suspicions of contagion continue, I commend your remaining there and awaiting (as you say you wish to do) the safety assurance from your closest friends, who, if not with greater love, at least with more certainty than we possess, will be able to apprise you of the facts.

But meanwhile I should judge that it would be wise to draw a profit from the wine in your cellar, at least one cask's worth; because although for now it is keeping well, I fear this heat may precipitate some peculiar effect: and already the cask that you had tapped before you left, Sire, from which the housemaid and the servant drink, has begun to spoil. You will need to give orders as to what you want done, because I have so little knowledge of this business; but I am coming to the conclusion that since you produced enough to last the entire year, and as you have been away for six of those months, you will still have plenty left, even if you should return in a few days.

Leaving this aside, however, and turning to that which concerns me more, I am longing to know in what manner your affair was terminated to the satisfaction of both you and your adversaries, as you intimated in the next to last letter you wrote me from Rome: tell me the details at your convenience, and only after you have rested, because I can be patient a while longer awaiting enlightenment on this contradiction.

Signor Geri was here one morning, during the time we suspected you to be in the greatest danger, Sire, and he and Signor Aggiunti went to your house and did what had to be done, which you later told me was your idea, seeming to me at the time well conceived and essential, to avoid some worse disaster that might yet befall you, wherefore I knew not how to refuse him the keys and the freedom to do what he intended, seeing his tremendous zeal in serving your interests, Sire.

Last Saturday I wrote to Her Ladyship the Ambassadress with all the great love that I felt, and if I receive an answer, I shall share it with you. I close here because sleep assails me now at the third hour of the night, on which account you will excuse me, Sire, in the event I have said anything inappropriate. I return to you doubled all the regards you offered to those named in your letter and especially La Piera and Geppo, who are thoroughly cheered by your return; and I pray blessed God to give you His holy grace.

FROM SAN MATTEO, THE 13TH DAY OF JULY 1633.
Most affectionate daughter,

S. M. Celeste

The deed that Signor Geri Bocchineri accomplished in Galileo's absence with the help of their young mutual friend, Niccolò Aggiunti, head of mathematics at Pisa, involved the destruction of potentially incriminating evidence. Fearing that representatives of the Inquisition might requisition Galileo's papers, his supporters took the prudent path of editing what could be found on his property.

Even though Signor Geri arrived at the convent in advance of Galileo's letter announcing his mission, Suor Maria Celeste surrendered her keys to him without hesitation. Geri served as a trusted link to her father throughout this period, and no one but her father could have informed him that she held the needed keys in safekeeping. These must have opened a certain closet or cabinet in Galileo's study. For surely Geri could have obtained keys to the gate and doors of the villa from Signor Rondinelli, who enjoyed free access to Il Gioiello, or from La Piera, the housekeeper, who would have opened any door to his familiar face – he being, as Sestilia's brother, practically a member of the family. But only Suor Maria Celeste could help him unlock the trove of private papers.

Her description of the event shows she did not judge Signor Geri's action inappropriate, nor herself an accomplice to any crime against the Holy Office. Apparently nothing she – or Geri and Aggiunti, or Galileo – had done in this connection required asking

God's forgiveness, or she would have shared her prayer for it here in writing, in her customary way.

Her confusion about the contradictory outcome of Galileo's affairs in Rome – things appearing to her to have been settled to the satisfaction of both sides – stemmed from letters he sent just before being sentenced. At the end of May, as shown in surviving correspondence with his friends, Galileo fully expected leniency for himself and the saving of face for all concerned, though in the end nothing had turned out that way.

'I do not hope for any relief,' Galileo wrote to a former pupil in France after the trial, 'and that is because I have committed no crime. I might hope for and obtain pardon, if I had erred; for it is to faults that the prince can bring indulgence, whereas against one wrongfully sentenced while he was innocent, it is expedient, in order to put up a show of strict lawfulness, to uphold rigour.'

Though the sisters of San Matteo had been overjoyed to hail Galileo's 'happy return' to Tuscany from Rome, it soon occurred to them how far away from home Siena really lay. Not just the day's ride over forty miles of hilly terrain still barred the way of reunion, but the pope's anger as well. Galileo had not yet been granted permission to return to Arcetri. In fact, Urban might never allow him to go home.

'When you were in Rome, Sire,' Suor Maria Celeste wrote to Galileo on 16 July, 'I said to myself: if I have the grace of your leaving that place and coming as far as Siena I will be satisfied, for then I can almost say that you are in your own house. And now I am not content, but find myself longing to again have you here even closer. Be that as it may, blessed be the Lord for having granted us His grace so magnanimously until now. It falls to us to try to be truly grateful for this much, so that He may be the more favourably disposed and compassionately moved to bless us in other ways in the future, as I hope He will do by His mercy.'

In Rome, Ambassador Niccolini continued to press for Galileo's return to Il Gioiello and notified him of each minuscule development in this campaign as a way of keeping hope alive.

*M*OST ILLUSTRIOUS AND BELOVED LORD FATHER

I READ THE LETTER you wrote to Signor Geri with particular pleasure and consolation, Sire, on account of the things contained in its first paragraph. I will be so bold as to venture on into the third paragraph as well, although it pertains to the purchase of some little house I do not know about, which I have inferred that Signor Geri very much wants Vincenzio to buy, albeit with your help. I certainly would not want to be presumptuous, interfering in matters that do not concern me. Nevertheless, because I care a great deal about whatsoever is of even minimal interest to you, Sire, I would implore you and exhort you (assuming you are in a position to be able to do this) to give them, if not the full amount, then some appreciable part of it, not only for love of Vincenzio, but just as much to keep Signor Geri favourably disposed towards you, as he has, on past occasions, shown great fondness for you, Sire, and, from all I have seen, tried to help you in any way he could: therefore, if, without too much trouble on your part, you could give him some sign of gratitude, I should judge that a deed well done. I know that you yourself can perceive and arrange such matters infinitely better than I, and perhaps I do not even know what I am saying, but well I know how anything I say is dictated by pure love towards you.

The servant who was in Rome with you came here yesterday morning, urged to do so by Signor Giulio Nunci. It seemed strange to me not to see letters from you, Sire. Yet I was appeased by the excuse this same man made, explaining that you had not known whether he would pass this way. Now that you are without a servant, Sire, our Geppo, who cannot move freely about here, desires nothing more, if only he were granted permission, than to come to you, and I should very much like that, too. If your thoughts concur, Sire, I could see to sending him well escorted, and I believe Signor Geri can secure him a permit to travel.

I also want to know how much straw to buy for the little mule, because La Piera fears she will die of hunger, and the fodder is not good enough for her, as she is a most original animal.

Since I sent you the list of expenses paid out for your house, we have incurred these others that I give you account of now, besides the money that every month I have made sure was paid to Vincenzio Landucci, for which I keep all the receipts, except the last two payments; for at those times he was, as he continues to be now, locked up in his house with the two little children because the plague killed his wife; whereby truly one may say she is released from her toil and gone to her rest, the poor woman. He sent early to ask me for the 6 *scudi* for the love of God, saying they were dying of hunger, and as the month was almost at its end I sent him the money; he promised the receipt when he is beyond suspicion of contagion, and I will endeavour to hold him to that; if nothing else I will first see to these other disbursements, in the event you are not here to take care of them yourself, Sire, which I suspect on account of the excessive heat that is upon us.

The lemons that hung in the garden all dropped, the last few remaining ones were sold, and from the 2 *lire* they brought I had three masses said for you, Sire, on my own initiative.

I wrote to Her Ladyship the Ambassadress, as you told me to, and sent the letter to Signor Geri, but I do not have a reply, wherefore I suppose I might be wise to write again suggesting the possibility that either my letter or hers has gone astray. And here, sending you love with all my heart, I pray Our Lord to bless you.

FROM SAN MATTEO, THE 24TH DAY OF JULY 1633.
Your most affectionate daughter,

S. Maria Celeste

Vincenzio had used part of Sestilia's dowry to put the down payment on their Costa San Giorgio house years before. Galileo had also contributed his share then, for he was named as an owner on the deed. The house included a garden, a reservoir and a court-yard, but its rooms were few. Now the building immediately adjacent to it had come up for sale, presenting an irresistible opportunity to expand the young family's quarters without their having to move.

While Galileo considered this proposal, he began to improve his health and outlook by engaging his mind in a new puzzle: Arch-bishop Piccolomini put him to work on the problem of recasting the giant bell for the cathedral's campanile.

[XXVII]

Terrible destruction on the feast of San Lorenzo

SIENESE FOUNDRYMEN HAD ERECTED the mould for the new tower bell out on the street, at the foot of the tall campanile with its many-windowed tiers. The mould consisted of two clay parts – one to sculpt the outer bell curve and one the inner – nested and resting upside down inside a huge scaffold. In order to maintain the crucial spacing between the two halves, the workers suspended the great weight of the inner mould across the rim of the outer by beams, like a sieve over a teacup. But when they started to pour the molten metal, the mould's inner section mysteriously rose and wrecked the contours of the resulting bell. Much surprise and speculation bruited through the piazza before Galileo offered the correct solution, which he proposed to reveal through a demonstration in the archbishop's home.

He called for an exact wooden model of the inner half of the bell mould, and when it arrived he inverted it and filled it with shot to make it heavy. Then he placed the mould model inside a glass urinal. The chamber-pot cradled the bell mould, according to the archbishop's description, 'leaving between the glass and wood a space the thickness of a *piastra* [a heavy silver coin]'. Next Galileo began to pour mercury into the urinal through a hole near the top. As soon as the quicksilver climbed just a short way up the walls of the glass container, it lifted the shot-filled mould model – though the model weighed twenty times as much as the piddling amount of mercury underneath it. Galileo had predicted this effect on the basis of his early experiments with floating bodies, wherein he showed how even a small child might lift a heavy load merely by pouring a little water.

The same effect had undermined the casting of the bell, Galileo claimed: the liquified metal had quickly set the interior mould afloat, despite its great weight. On the next attempt, he counselled, the workers must tightly anchor the top handles of the inner mould to the pavement to prevent a recurrence.

'And thus,' the archbishop was pleased to observe, 'the second time, the casting went very well.'

$\mathcal{M}$OST BELOVED LORD FATHER

IF MY LETTERS, AS you told me in one of yours, often reach you coupled in pairs, then I can tell you, not to repeat your exact words, that in this last post your letters arrived like the Franciscan friars wearing their wooden clogs, not only yoked together, but with a resounding clatter, creating in me a much greater than usual commotion of pleasure and happiness, Sire, especially when I learned that my supplication on behalf of Vincenzio and Signor Geri, which I submitted to you, or rather urged upon you, to speak more accurately, has been agreed to and settled so promptly and with even more generosity than I had requested: and consequently I conclude that my importuning in no way posed a disturbance to your peace, for indeed that possibility had worried me greatly, and now I feel cheered and relieved and I thank you.

As for your return, God knows how much I desire it; nevertheless, Sire, when you consider taking your leave from that city, where it has suited you for some time to remain in a place quite nearby, yet outside your own house, I should deem it better for both your health and your reputation, to stay on for several more advantageous weeks where for now you inhabit a veritable paradise of delights, especially considering the enchanting conversation of that most illustrious Monsignor Archbishop; rather than to have to return right away to your hovel, which has truly lamented your long absence; and particularly the wine casks, which, envying the praise you have lavished on the vintages of those other regions, have taken their revenge, for one of them has spoiled its contents, or indeed the wine has contrived to spoil itself, as I have already warned you might happen. And the other would have done the same, had it not been prevented by the shrewdness and diligence of Signor Rondinelli, who by recognising the malady has prescribed the remedy, advising and working to bring about the sale of the wine, which has been accomplished, through Matteo the merchant, to an innkeeper. Just today two mule loads are being decanted and sent off, with Signor Rondinelli's assistance. These sales, I believe, must bring in 8 *scudi*: any surplus left over after the two loads will be bottled for the family and the convent as we will gladly take this little bit: it seemed imperative to seize such an expedient before the wine sprang any other surprise on us that would have necessitated throwing it away. Signor Rondinelli atttributes the whole misfortune to our not having separated the liquid from the sediment in the casks before the onset of the hot weather; something I did not know about, because I am inexperienced in this enterprise.

The grapes in the vineyard already looked frightfully scarce before two violent hail-storms struck and completed their ruination. A few grapes were gathered in the heat of July before the arrival here of the highwaymen, who, not finding anything else to steal, helped themselves to some apples. On the feast day of San Lorenzo there came a terribly destructive storm that raged all around these parts with winds so fierce that they wreaked great havoc, and touched your house as well, Sire, carrying away quite a large piece of the roof on the side facing Signor Chellini's property, and also knocking over one of those terracotta flower pots that held an orange tree. The tree is transplanted in the ground for the time being, until we have word from you as to whether you want another pot purchased to hold it, and we reported the roof damage to the Bini family [the now-deceased Signor Martellini's in-laws], who promised to have it repaired.

The other fruit trees have borne practically nothing; particularly the plums, of which we had not a single specimen; and as for those few pears that were there, they have been harvested by the wind. However, the broad beans gave a very good yield, which, according to La Piera, will amount to 5 *staia* [less than a bushel] and all of them beautiful: now we must see to the white beans.

It would behoove me to give you an answer concerning your inquiry about whether or not I sit idle; but I am saving that until some time when I cannot sleep, as it is now the third hour of the night. I send you greetings on behalf of everyone I have mentioned, and even more from Doctor Ronconi who never comes here without pressing me for news of you, Sire. May the Lord God bless you.

FROM SAN MATTEO, THE 13TH DAY OF AUGUST 1633.
Most affectionate daughter,

S. M. Celeste

Even in common parlance in a Catholic country, any day of the month could be communicated by its religious significance as easily as by its number. Thus Galileo understood the freak storm that rent his roof and toppled an orange tree to have occurred on 10 August – from the mention of San Lorenzo, who met his martyrdom bound to a red-hot gridiron, quipping through this torment that his executioners should turn him over, for he had cooked enough on one side.

The stultifying heat of the summer of 1633 oppressed the village of Arcetri. Even Suor Maria Celeste complained about the weather, for the heat always weakened her. Unable to sleep, she said she could barely find the strength to move her pen. This was of course an exaggeration, for she wrote her father a minimum of two letters each week, answering his call for all the minutiae of home. Occasionally she gave him news of the convent, too – of how Suor Giulia, for example, at the age of eighty-five, had locked arms with Death and won. But she said little of her own un-flagging attention to the ailing sisters in the infirmary, or how her omnicompetence positioned her as a likely candidate to become mother abbess. She might have been elected the previous December, after she had reached the proper age for holding office, but she had been strapped with her father's concerns then, on the verge of his forced departure for Rome. Perhaps by the time of the next election, in 1635, the other nuns would look to her for the sort of leadership to be expected from one so intelligent and caring.

Regarding the sick Sisters, the Abbess is strictly bound by
herself and through other Sisters to inquire with solicitous
concern about what their illness requires both in the way of
counsel as also in foods and other necessary things, and so
provide tenderly and compassionately according to the possi-
bility of the situation. For all are bound to provide for and
serve their sick Sisters as they themselves would wish to be
served if struck down by any illness.

THE RULE OF SAINT CLARE, chapter VIII

That summer she also doctored Galileo's servant boy, Geppo,
who had spent a few days in the Florence hospital on account of
a feverish illness involving his spleen. Although his youth and
strength saw him quickly through this crisis, he was discharged
with a disgusting-looking skin disease acquired from some fellow
patient. Suor Maria Celeste cured it with an ointment of her own
preparation.

Later on in the unremitting heat, the servants ran out of flour.
But since there was no question of lighting the oven anyway that
August, Geppo bought bread for La Piera and himself from the
convent store. It cost only eight *quattrini* for a large loaf, as Suor
Maria Celeste informed Galileo, well in the habit now of keeping
his accounts.

Despite the heat, Galileo flourished in the favourable emotional
climate at Siena. Presently he resumed work on the book he had
been meaning to write for at least twenty-five years.

He had first set down a preliminary treatise on motion while a
professor at Pisa, but never published it. Then he laid new experimental groundwork for it during his two decades in Padua, where
he measured the swinging of pendulums until he could describe
their periods by a mathematical law, and where he rolled bronze
balls down inclined planes a thousand ways to derive the rate of
acceleration in free fall – in whatever time he could spare between
meeting teaching obligations and running a cottage industry in
military compasses. Later, as court philosopher at Florence in 1618,
with the promise of more leisure for such pursuits, he reopened
the labelled folders of his Paduan notes – only to be waylaid first

by illness and then by comets. He returned once more to the project early in 1631, while awaiting permission to publish the *Dialogue*. Now, detained at the archbishop's palace, Galileo revisited his ideas about the way everyday objects move, bend, break and fall.

'There is perhaps nothing in Nature older than MOTION', Galileo noted of the humdrum topic for his next book, 'about which volumes neither few nor small have been written by philosophers.' But all of those earlier texts had concerned themselves with pinning down the cause of motion. Galileo proposed to strike out on a different course – to drop all Aristotelian talk of *why* things moved, and focus instead on the *how*, through painstaking observations and measurements. In this fashion, he had discovered and described phenomena that generations of earlier philosophers had not even noticed. For example, the shape of the path traced through space by a hurled or fired missile, Galileo showed, was not just 'a line somehow curved', as his predecessors had said, but always precisely a parabola. And when lemons dropped from treetops, or cannonballs from towers, each one picked up speed in the same characteristic pattern tied to the elapsed time of its fall: whatever distance the object covered in one instant – measured as a pulse beat, a sung note, the weight of water that dripped from Galileo's timing device – by the end of two such instants it would travel four times as far. After three instants, it wound up at nine times the initial distance of descent; after four instants, sixteen distance units – and so on, always accelerating, always arriving at a distance determined by the square of the time passed.

Aristotle had ruled out any such mathematical approach to physics, on the grounds that mathematicians pondered immaterial concepts, while Nature consisted entirely of matter. And Nature, furthermore, could not be expected to follow precise numerical rules.

Galileo argued against this stance: 'Just as the accountant who wants his calculations to deal with sugar, silk and wool must discount the boxes, bales and other packings, so the mathematical scientist, when he wants to recognise in the concrete the effects

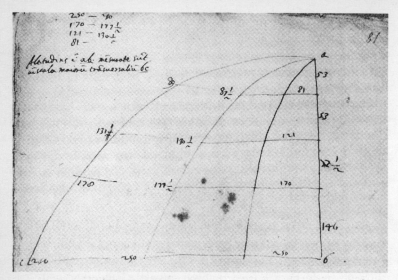

Manuscript page from Galileo's book *Two New Sciences*

he has proved in the abstract, must deduct any material hindrances [such as friction or air resistance]; and if he is able to do that, I assure you that things are in no less agreement than are arithmetical computations. The trouble lies, then, not in abstractness or concreteness, but with the accountant who does not know how to balance his books.'

Galileo envisioned the experimental, mathematical analysis of Nature as the wave of the future: 'There will be opened a gateway and a road to a large and excellent science', he predicted, 'into which minds more piercing than mine shall penetrate to recesses still deeper.'*

While Galileo devoted his time at Siena to writing, Ambassador Niccolini in Rome tirelessly pursued his full repatriation. The pope, however, would not be pressured into a promise, thus leaving the final sentence an open question. Rumours spoke of

---

* Galileo's last book, *Discourses and Mathematical Demonstrations Concerning Two New Sciences*, indeed ignited later physicists: Sir Isaac Newton transformed Galileo's ideas into laws of motion and universal gravitation.

Galileo's possible confinement after Siena at the Certosa, a vast hilltop monastery built in the fourteenth century to the south of Florence, where the twelve resident monks produced a locally famous wine. Such a move would bring Galileo even closer to Arcetri, facilitating the exchange of letters with his daughter, while ruling out any chance of his seeing her.

MOST BELOVED LORD FATHER

WHEN I WROTE TO YOU about your coming home soon, Sire, or your otherwise remaining where you are for a while longer, I knew of the petition you had made to his lordship the Ambassador, but was not yet aware of his answer, which I since learned from Signor Geri when he came here last Tuesday, just after I had written yet another letter to you, enclosing the formulation of the pills that by now must surely have reached you. My motive for addressing you in that seemingly distant fashion had grown out of my frequent discussions with Signor Rondinelli, who all through this period has been my refuge (because, as practical and experienced as he is in the ways of the world, he has many times alleviated my anxiety, prognosticating for me the outcome of situations concerning your affairs, especially in cases that seemed more precipitous to me than they later turned out to be); once during those discussions he told me how people in Florence were saying that when you departed from Siena, Sire, you would have to go to the Certosa, a condition that displeased every one of your friends; yet he saw some good in going along with those orders, as I understand the Ambassador himself did, too, for they both suspected that soliciting too urgently for your direct return here, Sire, might bring about some negative consequence, and therefore they wanted to allow more time to elapse before entreating again. Whereupon I, fearing the worst could all too easily come to pass, and hearing you were preparing to petition yet again, set myself to write to you as I did.

If ever I fail to make a great demonstration of the desire I harbour for your return, I refrain only to avoid goading you too much or disquieting you excessively. Rather than take that risk, all through these days I have been building castles in the air, thinking to myself, if, after these two months of delay in not obtaining the favour of your release, I had been able to appeal to Her Ladyship the Ambassadress, then she, working through the sister-in-law of His Holiness, might have successfully implored the Pope on your behalf. I know, as I freely admit to you, that these are poorly drawn plans, yet still I would not rule out the possibility that the prayers of a pious daughter could outweigh even the protection of great personages. While I was wandering lost in these schemes, and I saw in your letter, Sire, how you imply that one of the things that fans my desire for your return is the anticipation of seeing myself delighted by a certain present you are bringing, oh! I can tell you that I turned truly angry; but enraged in the way that blessed King David exhorts us in his psalm where he says, *Irascimini et nolite peccare* [Be angry, but sin not]. Because it seems almost as though you are inclined to believe, Sire, that the sight of the gift might mean more to me than that of you yourself: which differs as greatly from my true feelings as the darkness from the light. It could be that I mistook the sense of your words, and with this likelihood I calm myself, because if you questioned my love I would not know what to say or do. Enough, Sire, but do realise that if you are allowed to come back here to your hovel, you could not possibly find it more derelict than it is, especially now that the time approaches to refill the casks, which, as punishment for the evil they committed in allowing the wine to spoil, have been hauled up onto the porch and there staved in according to the sentence pronounced on them by the most expert wine drinkers in these parts, who point out as the primary problem your practice, Sire, of never having broken them open before, and the same experts claim the casks cannot suffer now for having had some sunshine upon their planks.

I received 8 *scudi* from the sale of the wine, of which I spent 3 on six *staia* of wheat, so that, as the weather turns cooler, La Piera may return to her bread baking; La Piera sends her best regards to you, and says that if she were able to weigh your desire to return against her longing to see you, she feels certain her side of the scale would plummet to the depths while yours would fly up to the sky: of Geppo there is no news worthy of mention. Signor Rondinelli this week has paid the 6 *scudi* to Vincenzio Landucci and has retained two receipts, one for last month, one for this: I hear that Vincenzio and the children are healthy, but I do not know how they are getting along, not having been able to inquire after them from a single person.

I am sending you another batch of the same pills, and I greet you with all my heart together with our usual friends and Signor Rondinelli. May Our Lord bless you.

FROM SAN MATTEO, THE 20TH DAY OF AUGUST 1633.
Most affectionate daughter,

S. Maria Celeste

[XXVIII]

Recitation of the penitential psalms

GALILEO IN SIENA RECONVENED his three familiar characters – Salviati, Sagredo and Simplicio – and let them take up the discussion of his *Two New Sciences*. The banning of the *Dialogue* had destroyed his literary monument to the memory of those long-gone friends, and so, although Galileo had no idea whether he would be allowed to publish another book, he began to put his accumulated material on motion into their mouths. He composed the dialogue for the first two of his three interlocutors' four new days together during his five-month stay at the archbishop's house.

In the initial excitement of greeting one another again, the three companions flood their talk with speculations on how to measure the speed of light or the weight of air, and how patterns of waves create consonance or dissonance in music. But their voices have changed since last they met. It is as though – it must be *because* – all three have lived through Galileo's tribulations with him and lost their verve. Salviati is not as persuasive, Sagredo not as passionate, Simplicio not nearly as stubbornly opposed to novelty. In lieu of the sarcastic barbs and literary devices that animated the *Dialogue*, the men trade polite lines to maintain the dramatic form, but not the flare, of the previous book. On the third and fourth days, they open their Italian dialogue to Salviati's reading aloud an entire Latin treatise on motion written by 'our Academician', as they like to call Galileo. These sections, dense as any geometry textbook, almost immediately lose the casual reader in a forest of propositions and theorems, though they embody Galileo's seminal restructuring of physics as a science based on mathematics.

Instead of Sagredo's palazzo on the Grand Canal, Galileo staged the trio's reunion at the great shipworks of the Venice Arsenale, where they could draw inspiration from the sight of men and machines at work. 'The constant activity which you Venetians display in your famous Arsenale', Salviati exclaims in his opening speech, 'suggests to the studious mind a large field for investigation, especially that part of the work which involves mechanics.'

The Venetian Arsenale

An experienced artisan at the shipyard immediately sets them off with his remark that extra care must be taken in the launching of the largest vessels, so as to avoid the peril of the big ships' splitting apart under their own great weight. Sagredo thinks this explanation strains belief. Though it is 'proverbial and commonly accepted,' he says, that large structures are weaker than small, 'I hold it to be altogether false, like many another saying which is current among the ignorant.'

Sagredo's doubt gives Salviati a chance to reveal the wisdom of the old workman's words, supported by 'our friend' Galileo's mathematical demonstrations regarding relative size and strength of materials – the first of the two 'new sciences' in the book's title. (The second, motion, including free fall and the paths of fired projectiles, would be covered on Days Three and Four.)

'Please observe, gentlemen,' Salviati responds,

> how facts which at first seem improbable will, even on scant explanation, drop the cloak which has hidden them and stand forth in naked and simple beauty. Who does not know that a horse falling from a height of three or four *braccia* will break his bones, while a dog falling from the same height or a cat from eight or ten, or even more, will suffer no injury? Equally harmless would be the fall of a grasshopper from a tower or the fall of an ant from the distance of the Moon. Do not children fall with impunity from heights which would cost their elders a broken leg or perhaps a fractured skull? And just as smaller animals are proportionately stronger and more robust than the larger, so also smaller plants are able to stand up better than larger. I am certain you both know that an oak two hundred feet high would not be able to sustain its own branches if they were distributed as in a tree of ordinary size; and that Nature cannot produce a horse as large as twenty ordinary horses, or a giant ten times taller than an ordinary man, unless by miracle or by greatly altering the proportions of his limbs and especially of his bones, which would have to be considerably enlarged over the ordinary. Likewise the current belief that, in the case of artificial machines the very large and the small are equally feasible and lasting is a manifest error.

The diagrams and geometrical proofs that follow show how volume outstrips strength as things get bigger: volume expands as the cube of bodies' dimensions, while strength increases only as much as their square.

Galileo's comparative bone drawing from *Two New Sciences*

'I am quite satisfied', declares Simplicio towards the end of the first day, 'and you may both believe me that if I were to begin my studies over again, I should try to follow the advice of Plato and commence from mathematics, which proceeds so carefully, and does not admit as certain anything except what it has conclusively proved.'

On Day Two, when the discussion becomes more mathematical and even more dependent upon diagrams, Salviati elaborates on problems of scale by drawing a couple of bones. One appears to be a femur from a dog. The other looks like a gross, bloated distortion of the same. 'To illustrate briefly,' Salviati says,

I have sketched a bone whose natural length has been increased three times and whose thickness has been multiplied until, for a correspondingly large animal, it would perform the same function which the small bone performs for its small animal. From the figures here shown you can see how out of proportion the enlarged bone appears. Clearly then if one wishes to maintain in a great giant the same proportion of limb as that found in an ordinary man he must either find a harder and stronger material for making the bones, or he must admit a diminution of strength in comparison with men of medium stature; for if his height be increased inordinately he will fall and be crushed under his own weight. Whereas, if the size of a body be diminished, the strength of that body is not diminished in the same proportion; indeed the smaller the body the greater its relative strength. Thus a small dog could probably carry on his back two or three dogs of his own size; but I believe that a horse could not carry even one horse of his own size.

'I am delighted to hear of your good health and peace of mind', Suor Maria Celeste wrote in response to one of Galileo's progress reports on *Two New Sciences*, 'and that your pursuits are so well suited to your tastes, as your current writing seems to be, but for love of God may these new subjects not chance to meet the same luck as past ones, already written.'

Because of the subjects he had treated in his *Dialogue*, Galileo was now expected to perform penance as part of the process of contrition. The Holy Office had enjoined him to recite the seven penitential psalms once a week for three years, to aid his rehabilitation. Saint Augustine, around the dawn of the fifth century, had selected these particular psalms for daily study and prayer in trying times, as a way to guard and greaten one's faith.

> *O Lord, rebuke me not in thine anger, neither chasten me in thy hot displeasure.*
> *Have mercy upon me, O Lord; for I am weak: O Lord, heal me; for my bones are vexed.*
>
> PSALM 6: 1–2

The sacrament of penance, which the Protestants had rejected during the Reformation, increased in importance in seventeenth-century Italy after the Council of Trent. The penitent was required to reconcile himself with God and the Church by performing three kinds of acts: contrition, confession and satisfaction. Galileo had already expressed his contrition and confessed publicly by abjuration. He most likely confessed in private and in confidence to an individual priest as well, although there is no evidence – of necessity there would be no evidence – that he did so. Even though the council's decrees stipulated only a single confession annually, at Eastertide, a new spiritual emphasis on the introspective examination of conscience compelled many Catholics to confess their sins as often as once a month.

Satisfaction, the third act of the sacrament of penance, consisted in performing three classes of good works, namely prayer, fasting and the giving of alms. The recitation of the penitential psalms, in partial fulfilment of the prayer obligation, would have taken Galileo approximately one-quarter of an hour per week, on his knees.

> *Blessed is he whose transgression is forgiven, whose sin is covered.*
> *Blessed is the man unto whom the Lord imputeth not iniquity, and in whose spirit there is no guile.*
>
> PSALM 32: 1–2

September saw the end of the plague epidemic that had menaced life in Tuscany for two whole years. Grand Duke Ferdinando attributed the respite to the May procession of the Miraculous Madonna of Impruneta. He and his grandmother, Grand Duchess Cristina, ordered their finest craftsmen to create ornate tokens of appreciation – including a cross in carved rock crystal with gold decorative bands, fifteen silver votive vases, and a silver reliquary containing the skull of Saint Sixtus – which the Medici family sent in a grateful outpouring to the small church that housed the Virgin's holy image.

> *I acknowledged my sin unto thee, and mine iniquity have I not hid. I said, I will confess my transgressions unto the Lord; and thou forgavest the iniquity of my sin.*
>
> PSALM 32: 5

At the Convent of San Matteo in Arcetri, the disappearance of the plague coincided with another remarkable stroke of good fortune: through a series of deaths attributed to old age, the late brother of Suor Clarice Burci bequeathed the nuns a farm at Ambrogiana valued at more than five thousand *scudi*. In reporting this event to Galileo, Suor Maria Celeste estimated the coming year's harvest to yield 290 bushels of wheat, 50 barrels of wine, and 70 sacks of millet and other grains, 'so that my convent will be greatly relieved'. She anticipated that Galileo, too, would be relieved of constant requests for money, given the sisters' sudden affluence.

Their benefactor, well knowing the nuns could not leave the convent to tend crops and feed animals, had thoughtfully willed them a full complement of field hands and caretakers to remain on the property. Along with this largess, the Poor Clares of San Matteo inherited the responsibility of celebrating mass every day for four hundred years to pray for the immortal soul of Suor Clarice's brother. They also stood obliged to perform the Office for the Dead in his honour three times per year for the next two centuries.

To these requisite prayers, Suor Maria Celeste voluntarily added the psalms for her father's penance.

> *Forsake me not, O Lord: O my God, be not far from me.*
> *Make haste to help me, O Lord my salvation.*
>
> PSALM 38: 21–2

*M*OST BELOVED LORD FATHER

SATURDAY I WROTE TO YOU, Sire, and Sunday, thanks to Signor [Niccolò] Gherardini [a young admirer, and later biographer, of Galileo, who was related to Suor Elisabetta], your letter was delivered to me, through which, learning of the hope you hold out for your return, I am consoled, as every hour seems a thousand years to me while I await that promised day when I shall see you again; and hearing that you continue to enjoy your well-being only doubles my desire to experience the manifold happiness and satisfaction that will come from watching you return to your own home and moreover in good health.

I would surely not want you to doubt my devotion, for at no time do I ever leave off commending you with all my soul to blessed God, because you fill my heart, Sire, and nothing matters more to me than your spiritual and physical well-being. And to give you some tangible proof of this concern, I tell you that I succeeded in obtaining permission to view your sentence, the reading of which, though on the one hand it grieved me wretchedly, on the other hand it thrilled me to have seen it and found in it a means of being able to do you good, Sire, in some very small way; that is by taking upon myself the obligation you have to recite one time each week the seven psalms, and I have already begun to fulfil this requirement and to do so with great zest, first because I believe that prayer accompanied by the claim of obedience to Holy Church is effective, and then, too, to relieve you of this care. Therefore had I been able to substitute myself in the rest of your punishment, most willingly would I elect a prison even straiter than this one in which I dwell, if by so doing I could set you at liberty. Now we have come this far, and the many favours we have already received give us hope of having still others bestowed on us, provided that our faith is accompanied by good works, for, as you know better than I, Sire, *fides sine operibus mortua est* [faith without works is lifeless].

My dear Suor Luisa continues to fare badly, and because of the pains and spasm that afflict her right side, from the shoulder to the hip, she can hardly bear to stay in bed, but sits up on a chair day and night: the doctor told me the last time he came to visit her that he suspected she had an ulcer in her kidney, and that if this were her problem it would be incurable; the worst thing of all for me is to see her suffer without being able to help her at all, because my remedies bring her no relief.

Yesterday they put the funnels in the six barrels of rose wine, and all that remains now is to refill the cask. Signor Rondinelli was there, just as he also attended the harvesting of the grapes, and told me that the must was fermenting vigorously so that he hoped it would turn out well, though there is not a lot of it;

I do not yet know exactly how much. This is all that for now in great haste I am able to tell you. I send you loving regards on behalf of our usual friends, and pray the Lord to bless you.

FROM SAN MATTEO IN ARCETRI, THE 3RD DAY OF OCTOBER 1633. Most affectionate daughter,

S. M. Celeste

Have mercy upon me, O God, according to thy loving kindness: according unto the multitude of thy tender mercies blot out my transgressions.
Wash me thoroughly from mine iniquity, and cleanse me from my sin.
For I acknowledge my transgressions: and my sin is ever before me.

PSALM 51: 1–3

It is not known whether Galileo himself recited the prayers of his penance, either before or after Suor Maria Celeste assumed the burden, for this was a duty performed in private. In public, Galileo remained ever consistent in his conviction that he had committed no crime.

'I have two sources of perpetual comfort,' he wrote retrospectively to his French supporter Nicolas-Claude Fabri de Peiresc, 'first, that in my writings there cannot be found the faintest shadow of irreverence towards the Holy Church; and second, the testimony of my own conscience, which only I and God in Heaven thoroughly know. And He knows that in this cause for which I suffer, though many might have spoken with more learning, none, not even the ancient Fathers, has spoken with more piety or with greater zeal for the Church than I.'

Hear my prayer, O Lord, and let my cry come unto thee.
Of old hast thou laid the foundation of the earth: and the
heavens are the work of thy hands.
They shall perish, but thou shalt endure: yea, all of them shall
wax old like a garment; as a vesture shalt thou change them,
and they shall be changed:
But thou art the same, and thy years shall have no end.

PSALM 102: I, 25–7

[XXIX]

The book of life, or, A prophet accepted in his own land

DURING THIS EPISODE OF anticipated healing at Siena, Galileo sank periodically into despondency. In October he confided to his daughter that he felt as though his name had been stricken from the roll call of the living. The condemnation by the Holy Office, so far exceeding the contumely he had come to expect in reaction to his work, branded him an outcast in his own eyes. At his worst moments, he despaired of ever re-establishing his reputation, of ever bringing the rest of his work to light. All his life he had attracted jealousy and criticism, sustaining blows dealt in such number and with such vehemence that he esteemed himself a magnet for malignity.

'May it please Blessed God that the final decree regarding your return does not postpone it longer than we hope,' Suor Maria Celeste wrote right back brightly on 15 October.

> But meanwhile I take endless pleasure in hearing how ardently Monsignor Archbishop perseveres in loving you and favouring you. Nor do I suspect in the slightest that you are crossed out, as you say, *de libro viventium*,* certainly not throughout most of the world, and not even in your own country: on the contrary it seems to me from what I hear that while you may have been eclipsed or erased very briefly, now you are restored and renewed, which is a thing that stupefies me, because I am well aware that ordinarily: *Nemo Propheta acceptus in patria sua*† (I fear that my wanting to use the Latin phrase has perhaps made me utter some barbarism). And surely, Sire, here at the convent you are also beloved and esteemed more than ever; for all this may the Lord God be praised, as He is the principal source of these graces, which I consider my own reward, and thus I have no other desire but to show gratitude for them, so that His Divine Majesty may continue to concede other graces to you, Sire, and to us as well, but above all your health and eternal blessing.

Everything Suor Maria Celeste intimated about Galileo's standing in the wide arena of the world was true. His former pupils still revered him, and elsewhere in Europe they spoke out against the injustice of his condemnation. His supporters included René Descartes in Holland, and astronomer Pierre Gassendi along with mathematicians Marin Mersenne and Pierre de Fermat in France. The French ambassador to Rome, François de Noailles, who had studied under Galileo at Padua, campaigned for his pardon, marching into Rome in 1633 in lavish display, at the head of a cavalcade

---

* From the book of the living.
† No prophet is accepted in his own land.

of silver-shod horses attended by liverymen in gold-embroidered coats.*

Churchmen, too, let it be known that Galileo had been wronged, though few protested as boldly as the archbishop of Siena. In Venice, for example, Galileo could still rely on the loyalty of Fra Fulgenzio Micanzio, theologian to the Venetian republic, whom he had met during his years in Padua. Micanzio had weathered previous papal storms, as in 1606, when Pope Paul V imposed the interdict against Venice, virtually suspending the celebration of Catholic life in that territory for a full year as a punishment for the republic's flouting of his authority. Micanzio had stood by his former superior, Galileo's good friend Fra Paolo Sarpi, throughout that ordeal and until Sarpi's death in 1623, when he succeeded him. Similarly, Micanzio would stand by Galileo now.

The truly noteworthy attentions of Archbishop Piccolomini, meanwhile, reached well beyond the palace where Galileo remained his charge, all the way to Galileo's daughters at the Convent of San Matteo in Arcetri. The monsignor sent them frequent gifts, including his most excellent wine, which was shared among all the nuns, either by the glass or in their soup. Thanks to the archbishop, Galileo could proffer Suor Maria Celeste treats she had never seen or imagined, such as the creamy white egg-shaped lumps of mozzarella cheese made from water buffalo's milk.

'Lord Father, I must inform you that I am a blockhead,' she admitted in response to this promised gift, 'indeed the biggest one in this part of Italy, because seeing how you wrote of sending me seven "Buffalo eggs", I believed them truly to be eggs, and planned to fry a huge omelette, convinced that such eggs would be very grand indeed, and in so doing I made a merry time for Suor Luisa, who laughed long and hard at my foolishness.'

Here she punned by calling herself a *bufala* – a word meaning both 'blockhead' and 'female buffalo'. When she learned in a

---

* Later, Galileo thanked him by dedicating *Two New Sciences* 'To the very illustrious nobleman, my Lord the Count de Noailles, Councillor to his Most Christian Majesty; Knight of the Holy Ghost; Field Marshal of the Armies,' et cetera, et cetera.

subsequent letter 'that Monsignor Archbishop was well aware of my gaffe regarding the buffalo eggs, I could not help but blush for shame, although on the other hand I am happy to have given you grounds for laughter and gladness, as it is with this motive I so often write to you of foolish things'.

Ten months into her wearisome separation from her father, she remained his faithful chronicler on all matters of possible interest, even when her headaches and toothaches forced her to be brief. 'Signor Rondinelli', she reported in mid-October, 'has not shown his face here for a fortnight, because, from what I hear, he is drowning in the little bit of wine he had put in two kegs that are turning bad and giving him great grief.'

Although Galileo never left the palace grounds, the archbishop facilitated his affairs, enabling Suor Maria Celeste to request specific items at her whim. 'I have always wanted to know how to make those Sienese cakes that everyone raves about; now that All Saints' Day [1 November] is approaching, you will have the occasion, Sire, to let me see them, I do not say "taste" them so as not to sound gluttonous: you are further obliged (because of the promise you made me) to send me some of that strong reddish linen yarn which I would like to use to start preparing some little Christmas gift for Galileino, whom I adore because Signor Geri tells me that, beyond being the namesake, the boy also has the spirit of his grandfather.'

She further urged Galileo to write 'two lines' to his ever concerned doctor, Giovanni Ronconi, whom she saw frequently in the infirmary these days. Suor Maria Celeste summoned Doctor Ronconi or one of his confrères to the convent whenever an illness appeared too grave for her to treat alone.

> *Those who are sick may lie on sacks filled with straw and*
> *may have feather pillows for their heads, and those who need*
> *woollen socks and mattresses may use them.*
>
> THE RULE OF SAINT CLARE, chapter VIII

Although the danger of the plague had passed, fevers and persistent malaises now afflicted five of the nuns.

MOST BELOVED LORD FATHER

LAST WEDNESDAY A BROTHER of the Priory of San Firenze came to bring me a letter from you along with the little package of russet linen yarn, which, considering the rather thick quality of the thread, seems somewhat expensive; but indeed the colour of the dye, being very beautiful, makes the price of six *crazie* per skein appear more tolerable.

Suor Luisa stays in bed with only the slightest improvement, and in addition to her, several of the others are also sick, so that if we now faced any suspicion of plague we would be lost. Among the sick is Suor Caterina Angela Anselmi who was formerly our Mother Abbess, a truly venerable and prudent nun, and, after Suor Luisa, the dearest and most intimate friend I have ever had: she is so gravely ill that yesterday morning she received the Extreme Unction, and it appears that she has only a few days to live; and the same must be said for Suor Maria Silvia Boscoli, a young woman of 22 years, and you may recall, Sire, how people once spoke of her as the most beautiful girl to grace the city of Florence for 300 years: this marks the sixth month she has been lying in bed with a continuous fever that the doctors now say has turned to consumption, and she is so wasted as to be unrecognisable; yet with all that she retains a vivacity and energy especially in her speech that astounds us, while from hour to hour we doubt whether that faint spirit (which seems entirely confined to her tongue) will fade away and abandon the already exhausted body: then, too, she is so listless that we can find no nourishment to suit her taste, or to say it better, that her stomach can accept, except a little soup made from broth in which we have boiled some dried wild asparagus, and these are extremely difficult to find at this time of year, wherefore I was thinking that perhaps she could take some soup made from grey partridge, which has no gamy taste. And since these birds abound where you are, Sire, as you say in your letters, you might be able to send me one of them for her and for Suor Luisa, and I doubt you would encounter any difficulty having them reach me in good condition, since our Suor Maria Maddalena Squadrini recently received several fresh, good thrushes that were sent by a brother of hers who is Prior at the Monastery of the Angels, in a part of the diocese very close to Siena. If, without too much trouble, Sire, you could help me make such a gift, now that the idea has whetted my appetite, I would be ever so grateful.

This time it devolves upon me to play the raven who bears bad tidings, as I must tell you that on the feast day of San Francesco [4 October], Goro, who worked for the Sertinis, died, and left a family in great distress, according to his wife who was here yesterday morning to beseech me that I must convey this news to you, Sire, and furthermore remind you of the promise you made to Goro himself and to Antonia his daughter, to give her a black woollen house-dress when she got married: now they are in dire straits, and Sunday, which will be tomorrow, she will say her vows in Church; and because Goro's wife has spent what little money she had, first on medicaments and then for his funeral, she is hard-pressed, and wants to know if you can do her this kindness: I promised that I would tell her your answer as soon as I heard from you, Sire.

I would not know how to make you realise the happiness I derive from learning that you continue conserving your health in spite of everything, except to say that I enjoy your good fortune more than my own, not only because I love you more than myself, but also because I can imagine that if I were oppressed by infirmity, or otherwise removed from the world it would matter little or nothing to anyone, since I am good for little or nothing, whereas in your case, Sire, the opposite holds true for a host of reasons, but especially (beyond the fact that you do so much good and are able to help so many others) because the great intellect and knowledge that the Lord God has given you enables you to serve Him and honour Him far more than I ever could, so that with this consideration I come round to cheer myself and take greater pleasure from your well-being than from mine.

Signor Rondinelli has allowed himself to be seen again now that his kegs have quieted down; he sends his greetings to you, Sire, and so does Doctor Ronconi.

I assure you that I am never vexed by boredom, Sire, but sooner by the hunger caused, I believe, if not by all the exercise I perform, then by the coldness of my stomach, which does not get the full complement of sleep it requires, since I have no time. I rely on the oxymel and the papal pills to make up this deficit. I only tell you this to excuse myself for the haphazard appearance of my letter, as I was compelled to put down and then take up my pen again more than once before I could complete it, and on that note I commend you to God.

FROM SAN MATTEO IN ARCETRI, THE 22ND DAY OF OCTOBER 1633.
Your most affectionate daughter,

S. M. Celeste

The enclosed conforms to the wish you expressed in your previous letter, Sire, that after having written to you I should write also to Her Ladyship the Ambassadress. I suspect that my numerous activities have sapped my energy leaving me little to give her; you will be able to look it over and make corrections, and do let me know if you send her the ivory crucifix.

I still cling to the hope that this week you will have some resolution regarding your release, and I am burning with desire to share in that news.

According to medieval and Renaissance medical theory, each of the Earth's four elements – earth, fire, air, water – had its correspondent humour in the human body: black bile, yellow bile, blood, phlegm. These in turn denoted specific organs – spleen, liver, heart, brain – and conveyed the qualities, respectively, of dry cold, dry heat, moist heat and moist cold. Suor Maria Celeste spoke in diagnostic terms when she referred to the 'coldness' of her stomach, meaning that it behaved sluggishly. Foods, remedies and common activities could likewise be classified and prescribed by the same four qualities, so that the asparagus offered in broth to the once-lovely Suor Maria Silvia was described by herbalists as warm and moist in the first degree. Partridges, being of moderate warmth, were recommended for convalescents.

Doctor Ronconi, with a university education in natural philosophy and more options at his disposal, could purge or bleed patients as necessary to restore the body's balance. Being a physician gave him higher social status than the so-called barber-surgeons, most of whom received vocational training outside the university setting. This same disdain also tainted surgeons from the finest schools. Galileo compared the snobbishness shown by physicians towards surgeons to the groundless prejudice that philosophers lorded over mathematicians. Having befriended and benefited from the ministrations of the famed anatomist Girolamo Fabrici of Acquapendente, Galileo used him to make a point in defence of geometry.

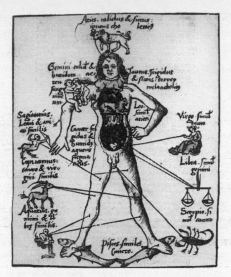

Sixteenth-century depiction of elements, humours and organs

I hear my adversaries shouting in my ears that it is one thing to deal with matters physically, and quite another to do so mathematically, and that geometers should stick to their fantasies and not get entangled in philosophical matters – as if truth could ever be more than one; as if geometry up to our time had prejudiced the acquisition of true philosophy; as if it were impossible to be a geometer as well as a philosopher – and we must infer as a necessary consequence that anyone who knows geometry cannot know physics, and cannot reason about and deal with physical matters physically! Consequences no less foolish than that of a certain physician who, moved by a fit of spleen, said that the great doctor Acquapendente, being a famed anatomist and surgeon, should content himself to remain among his scalpels and ointments without trying to effect cures by medicine – as if knowledge of surgery destroyed and opposed a knowledge of medicine. I replied to him that having many times recovered my health through the supreme excellence of Signor Acquapendente, I could depose and certify that he had never given me to drink

any compound of cerates, caustics, threads, bandages, probes
and razors, nor had he ever, instead of feeling my pulse, cau-
terised me or pulled a tooth from my mouth. Rather, as an
excellent physician, he purged me with manna, cassia or rhubarb, and used other remedies suitable to my ailments.

Well acquainted with her father's ailments, Suor Maria Celeste
regularly sent him her handmade papal pills, which contained
dried rhubarb (a natural laxative), saffron supplied from Siena by
Galileo, and aloe that had been washed with rose water no fewer
than seven times. Her sluggish digestive system relied on the same
medicine, although she took shortcuts in preparing her own pills,
leaving out whatever she lacked at the moment and settling for a
single rose water washing of the aloe. When caring for Arcangela
or another sick sister, however, she asked Galileo's help in procuring expensive ingredients such as Tettucio water (a superior purgative) and nutmeg oil to control nausea and vomiting. Such
substances, catalogued in the official Florentine *Farmacopoeia*,
could be formulated into pills, potions or powders following the
instructions in available medical manuals.

Suor Maria Celeste probably received her apprenticeship as an
apothecary under the nuns and visiting doctors who staffed the convent's infirmary. Her more basic schooling in letters and Latin, on
the other hand, undoubtedly came from her father, in whatever
time he could spare during her formative years, for it is clear that no
one at San Matteo surpassed her in language skills. Even the abbesses sought her out to write important letters of official business.

In Galileo's own scientific correspondence through the autumn
of 1633, he circulated among his friends certain proofs relating to
the strength of materials. With their permission, and to their
delight, he incorporated some of their additions and suggestions
into the text for the second day of *Two New Sciences*.

Galileo would later judge *Two New Sciences* 'superior to everything else of mine hitherto published', because its pages 'contain
results which I consider the most important of all my studies'. By
his own reckoning, then, his conclusions on resistance and motion

outweighed all the astronomical discoveries that immortalised his name. Surely Galileo prided himself on having been the first to build a proper telescope and point it towards the sky. But he believed his own greater genius lay in his ability to observe the world at hand, to understand the behaviour of its parts, and to describe these in terms of mathematical proportions.*

While he worked on the dialogue for the beginning of *Two New Sciences*, Galileo also wrote a play. He sent it to Suor Maria Celeste for performance by the nuns – apparently for the anticipated entertainment of Her Ladyship Caterina Niccolini, the wife of the Tuscan ambassador, who was still intent on visiting the convent. Unfortunately, nothing survives of Galileo's religious drama except his daughter's mention of it in thanks. 'The play, coming from you,' she wrote after reading the first act, 'can be nothing if not wonderful.'

In Rome, Ambassador Niccolini told Urban VIII that Galileo had proved himself a model prisoner in Siena by demonstrating obedience to the pope and the Holy Office. Urban weighed this claim against impeachments of Archbishop Piccolomini that reached him from clerics in Siena. It seemed that the archbishop quite often invited various scholars to his table, the better to enrich the intellectual repartee so beneficial to Signor Galilei's peace of mind. In other words, instead of holding Galileo prisoner as a confessed heretic, Piccolomini indulged him as a guest of honour.

'The Archbishop', an anonymous hand informed officials in Rome, 'has told many that Galileo was unjustly sentenced by this Holy Congregation, that he is the first man in the world, that he will live for ever in his writings, even if they are prohibited, and that he is followed by all the best modern minds. And since such seeds sown by a prelate might bear pernicious fruit, I hereby report them.'

---

* Posterity agrees with Galileo in this assessment of his merits. As Albert Einstein noted, 'Propositions arrived at purely by logical means are completely empty as regards reality. Because Galileo saw this, and particularly because he drummed it into the scientific world, he is the father of modern physics – indeed of modern science altogether.'

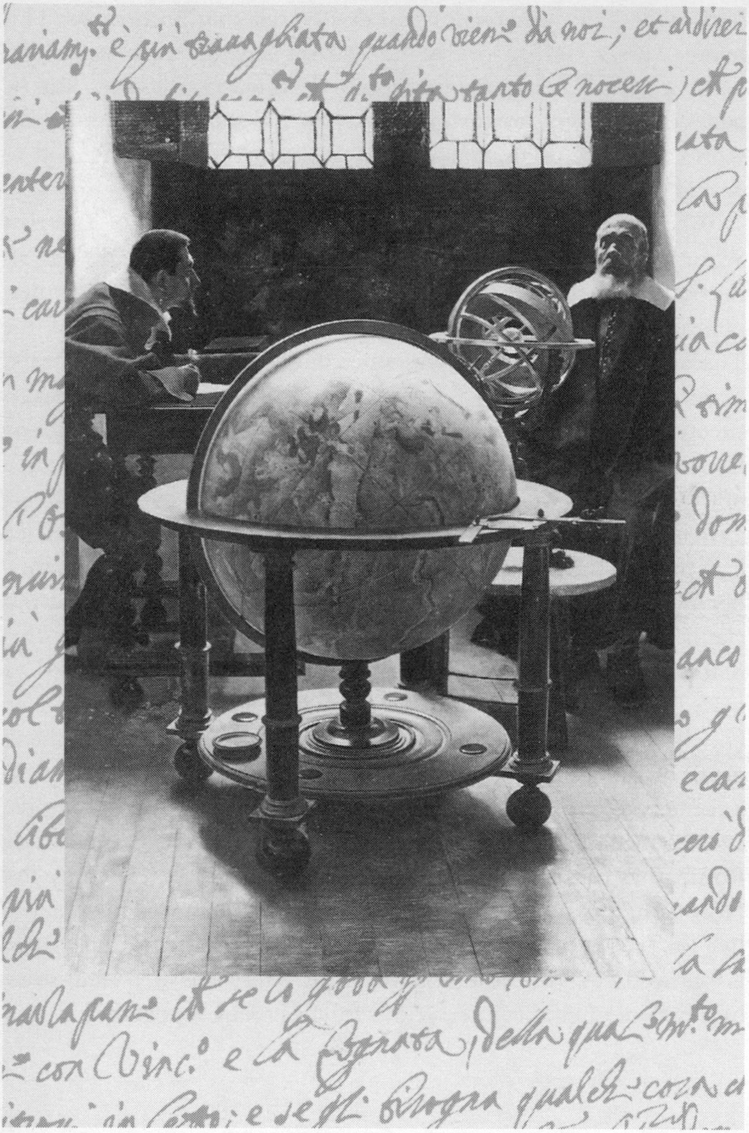

PART SIX

From Arcetri

[XXX]

My soul and its longing

IT RAINED ALL ACROSS Tuscany at the end of October 1633 and on into November. The dampness aggravated Galileo's arthritic pains, deepened Suor Maria Celeste's lassitude, cast its dreary pall over all their expectations. Suor Caterina Angela, the former mother abbess at San Matteo, died in the wet autumn weather, and the nuns buried her in the convent cemetery in the rain.

> *I am the resurrection, I am the life; he who believes in me, even if he die, shall live; and whoever lives and believes in me, shall never die.*

> OFFICE FOR THE DEAD, CANTICLE OF ZECHARIAH

Grey partridge

The other sick nuns in the infirmary held on. Galileo failed to send the partridge for them, though not through lack of trying. This late in the hunting season, not a single one could be bagged anywhere. Suor Maria Celeste, for her part, had no better luck procuring the tiny ortolan buntings that Galileo craved and could not get in Siena.

'I delayed writing this week', she apologised, 'because I really wanted to send you the ortolans, but in the end none have been found, and I hear they fly away when the thrushes arrive. If only I had known this desire of yours, Sire, several weeks ago, when I was racking my brain trying to think of what I could possibly send you that might please you; but never mind! You have been unlucky in the ortolans, just as I was foiled by the grey partridges, because I lost them to the goshawk falcon.'

The heavy rain made it impossible to set the broad beans in his garden, she said, but fair weather must come again, and with it his return. 'I send you no pills because desire makes me hope that you must soon arrive here to claim them in person: I am all eagerness to hear the resolution that will reach you this week.'

The resolution did not come, however. Galileo grew cranky with waiting, and so dependent on his daughter's letters that he scolded her when they failed to arrive often enough to soothe him.

'If only you could fathom my soul and its longing the way you penetrate the Heavens, Sire,' she began on 5 November, 'I feel certain you would not complain of me, as you did in your last letter; because you would see and assure yourself how much I should want, if only it were possible, to receive your letters every day and also to send you one every day, esteeming this the greatest satisfaction that I could give to and take from you, until it pleases God that we may once again delight in each other's presence.'

Then she told him how she had secured his coveted ortolans after all – through a bird-keeper in the service of the grand duke. Any moment now she would dispatch Geppo to the game-rich acres of the Boboli Gardens behind the Pitti Palace, armed with a floured box in which to pack the delicate birds and deliver them straight away to Signor Geri. But when Galileo thanked her for them a few days later, he said nothing of his own flight from Siena.

'I must tell you first how astounded I was', she wrote the following Saturday, 12 November, 'that you made no mention in your most recent letter of having received any word from Rome, nor any resolution regarding your return, which we had so hoped to have before All Saints' Day [1 November], from what Signor Gherardini led me to believe. I want you to tell me truly how this business is progressing, so as to quiet my mind, and also please tell me what subject you are writing about at present: provided it is something that I could understand, and you have no fear that I might gossip.'

Suor Maria Celeste necessarily understood the semi-secret nature of her father's works in progress. In fact, some of the materials removed from Il Gioiello during Galileo's trial concerned his writing on motion, including the manuscript for the third day of *Two New Sciences*, which he had roughed out before he left for Rome. Although Galileo would not have deemed these particular documents incriminating, he had reason to fear their needless destruction.

Day Three's treatment of uniform and accelerated motion bears witness to the untold number of Paduan hours Galileo spent tracking the course of a small bronze ball down the groove of an inclined plane to probe the mystery of acceleration. Unable to experiment fruitfully with freely falling objects, Galileo built his inclined-plane apparatus so he could control fall, stopping the action at will, making precise measurements of time and distance all the while. Salviati, who claims in *Two New Sciences* to have assisted Galileo occasionally in these efforts, describes them to Sagredo and Simplicio:

> We rolled the ball along the channel, noting, in a manner presently to be described, the time required to make the descent. We repeated this experiment more than once in order to measure the time with an accuracy such that the deviation between two observations never exceeded onetenth of a pulse-beat. Having performed this operation and having assured ourselves of its reliability, we now rolled the ball only one-quarter the length of the channel; and having measured the time of its descent, we found it precisely onehalf of the former. Next we tried other distances, comparing the time for the whole length with that for the half, or with that for two-thirds, or three-fourths, or indeed for any fraction; in such experiments, repeated a full hundred times, we always found that the spaces traversed were to each other as the squares of the times, and this was true for all inclinations of the plane . . . along which we rolled the ball.

Just as Copernicus had discerned the configuration of the solar system with no telescope to guide him, Galileo arrived at this fundamental relationship between distance and time without so much as a reliable unit of measure or an accurate clock. Italy possessed no national standards in the seventeenth century, leaving distances open to guesstimate gauging by fleas' eyes, hairbreadths, lentil or millet seed diameters, hand spans, arm lengths and the like. Even a *braccio* differed in dimension depending on whether it was measured in Florence, Rome or Venice, and so Galileo delineated his own arbitrary units along the length of his experimental apparatus. As long as these units matched one another, he could use them to establish fundamental relationships.

To clock the rolling time of the balls, Galileo literally weighed the moments. 'For the measurement of time', Salviati continues in his experimental description, 'we employed a large vessel of water placed in an elevated position; to the bottom of this vessel

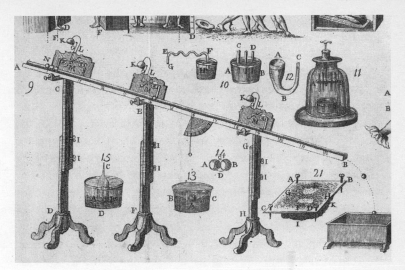

Artistic rendition of Galileo's incline plane

was soldered a pipe of small diameter giving a thin jet of water, which we collected in a small glass during the time of each descent, whether for the whole length of the channel or for a part of its length; the water thus collected was weighed, after each descent, on a very accurate balance [against grains of sand]; the differences and ratios of these weights gave us the differences and ratios of the times, and this with such accuracy that although the operation was repeated many, many times, there was no appreciable discrepancy in the results.'

Although this account reveals stunning experiments that promise to open a new window on philosophy, Salviati cannot be shaken from his recently acquired pedantic monotone, which threatens to establish an irreparable split, if not between science and religion, then between science and poetry.

The ball-and-plane trials provided the tedious yet triumphant prelude to the truth about falling, which Galileo expressed in *Two New Sciences* as a series of theorems. He did not use the convention of algebraic analysis that later allowed his rules to be pared down to a few letters and symbols, but expressed his findings as geometric ratios, and wrote out his proofs in dense prose accompanied by letter-labelled line drawings in the style of the ancient Greek mathematicians.

All the motions discussed on the third day of *Two New Sciences* are 'natural' ones, since the experimental objects simply roll or drop, instead of being thrown with force. Not until Day Four (which Galileo would complete several years later, in 1637) do the 'violent' movements of bullets and other projectiles come up for discussion. Here Galileo displays his singular insight in breaking motions into their separate components. For he shows that any cannonball fired from a mortar, for example, or any arrow shot from a bow, combines two vectors: the uniform forward thrust of the propulsion and the downward acceleration of free fall.

'One cannot deny that the argument is new, subtle and conclusive,' remarks Sagredo, 'resting as it does upon this hypothesis, namely, that the horizontal motion remains uniform, that the vertical motion continues to be accelerated downwards in proportion to the square of the time, and that such motions and velocities as these combine without altering, disturbing or hindering each other, so that as the motion proceeds the path of the projectile does not change into a different curve.'

Whatever the weight of the projectile or the force propelling it, Salviati explains on Galileo's say-so, the path traced in space always assumes the curved shape of a parabola. Shooting straight up or down, however, constitutes a separate category, also taken into consideration by the participants.

Mere mention of the parabola by Salviati instantly occasions a digression into the geometry of the cone, the parent body of the parabola, for the sake of Sagredo and Simplicio, who fear they may fail to follow a discussion promising such a plethora of parabolas. 'Your demonstration proceeds too rapidly', complains Simplicio, 'and, it seems to me, you keep on assuming that all of Euclid's theorems are as familiar and available to me as his first axioms, which is far from true.'

But once the shape has been thoroughly examined to the satisfaction of the less mathematically inclined, the discussion continues easily and amicably. It even allows for an analysis of the efficiency of various angles of elevation of heavy artillery, complete with a geometric proof showing why the forty-five-degree angle surpasses all others – because the parabola thus described has the greatest possible amplitude, and therefore the shot carries furthest.

'The force of rigid demonstrations such as occur only in mathematics fills me with wonder and delight,' Sagredo exclaims at this revelation. 'From accounts given by gunners, I was already aware of the fact that in the use of cannon and mortars, the maximum range, that is the one in which the shot goes furthest, is obtained when the elevation is 45° or, as they say, at the sixth point of the quadrant; but to understand why this happens far outweighs the mere information obtained by the testimony of others or even by repeated experiment.'

This emphasis on the practical application and value of science, so far removed from the metaphysical consideration of causes, set Galileo apart from most philosophers of his time. While Aristotelian philosophers talked of essences and natural places, Galileo went after quantifiable entities such as time, distance and acceleration. Other contemporary treatises in dialogue form typically situated the speakers around a university quadrangle. *Two New Sciences* takes place in a shipyard. Its characters simply dismiss the pursuit of ultimate causes that has passed for science until their time: 'The cause of the acceleration of the motion of falling bodies is not a necessary part of the investigation,' declares Salviati. From now on, physics will never be the same.

'The continuous rain', Suor Maria Celeste wrote in a morning-after postscript to her weekly summary of Saturday 12 November, 'has not allowed Giovanni (as the bearer of this letter is called) to leave this morning, which is Sunday, and this leaves me time to chat with you a little longer, and to tell you that recently I pulled a very large molar, which had rotted and was giving me great pain; but what is worse is that I have several others that soon will do the same.' She had begun extracting her own teeth years earlier – a self-treatment that must have required heroic determination –

and described herself as prematurely toothless when she was only twenty-seven.

'To respond to that personal detail you shared with me,' her postscript continued, 'that you find occupations so salubrious, truly I recognise them as having that same effect on myself as well: so that even though the activities occasionally seem superfluous and intolerable to me, on account of my being a friend of tranquillity, I nevertheless see clearly how staying active is the foundation of my health, and particularly in the time that you have been far away from us, Sire, with great providence did the Lord arrange it so that I never had what you might call an hour of peace, thus preventing the oppression of your absence from distressing me. Such grief would have been harmful to me, and given you cause for worry instead of the relief I have been able to provide.'

As Suor Maria Celeste sat writing in the Sunday morning rain of 13 November, Ambassador Niccolini again approached the pope for permission to send Galileo back to Arcetri. Urban withheld his judgment. He neither granted nor denied the request during this audience, but pointedly let on that he knew Galileo was enjoying the support, defence, companionship and correspondence of like-minded men – all of whom, Urban assured Niccolini with undisguised disgust, were now under surveillance by the Holy Office. The archbishop of Siena himself numbered among this company.

'If good luck had enabled you to find even one grey partridge,' Suor Maria Celeste filled the time writing as another week passed without news of her father's homecoming, 'I would have been thrilled to have it for love of that poor sick young girl, who craves nothing but wild game: at the last full moon she was so ill as to be anointed with holy oil, but now she has made such a comeback that we believe she will live to see the new moon. She speaks with great vivacity, and gulps her food readily, provided we give her tasty things. Last night I stayed with her all through the night, and while I fed her, she said: "I cannot believe that when one stands on the verge of death it is possible to eat the way I do, yet, for all that, I have no desire to turn back; only to see God's will be done."'

As many times as Suor Maria Celeste must have touched the back of her hand to Suor Maria Silvia's feverish forehead, she had no way to measure the girl's temperature.* She could gauge it only relatively, refining her reading by the flush of the patient's complexion, the rapidity of her pulse and the shallowness of her breathing.

Before the week was out, just as the daylight in the infirmary was fading one afternoon at the twenty-fourth hour, a messenger arrived from Siena with a pannier full of game birds.

Inside it were 12 thrushes: the additional 4, which would have completed the number you state in your letter, Sire, must have been liberated by some charming little kitten who thought of tasting them ahead of us, because they were not there, and the cloth cover had a large hole in it. How fortunate that the grey partridges and woodcocks were at the bottom, one of which and two thrushes I gave to the sick girl, to her great joy, and she thanks you, Sire. I sent another gift, also in the form of two thrushes, to Signor Rondinelli, and the remainder we enjoyed together with our friends.

I have taken the greatest pleasure in distributing all this bounty among various people, because prizes sought after with such diligence and difficulty deserve to be shared by several, and as the thrushes arrived a little the worse for wear, it was necessary to cook them in a stew, so that I stood over them all day, and for once I truly surrendered myself to gluttony.

* Galileo invented a rudimentary thermometer, around 1593, for approximating room temperature, but it took until 1714 for Daniel Fahrenheit to improve on the device by sealing mercury in glass and marking the tube with a degree scale calibrated by the freezing and boiling points of water.

[XXXI]

Until I have this from your lips

THE WEDNESDAY BEFORE THE start of Advent, when Suor Maria Celeste went searching for the pears Galileo had requested, she briefly imagined saving them to present to him in person – before her doubts dispelled these pleasant thoughts. 'Because I hear that this year the fruits do not last long, I wonder if it might be better, once I have them, to send them to you right away and not wait for your return, which could be delayed for several more weeks, or so my desire leads me to fear.'

By Church tradition, the hopeful time of Advent commenced with the Sunday closest to the feast day of Saint Andrew the Apostle, 30 November. In 1633, the nearest Sunday happened to fall on 27 November – the earliest possible date for Advent to begin, and this stretched the period to its maximum duration of twenty-eight days. Everything that had anything to do with waiting dilated during the dark season Suor Maria Celeste spent anticipating her father's return. In her Breviary she read the Advent Liturgy regarding 'the Lord the King that is to come', 'the Lord already near', and 'Him Whose glory will be seen on the morrow'.

Inside the tiny Church of San Matteo, Suor Barbera prepared the violet-coloured candles, the seasonal violet altar hangings, and the special violet and rose vestments for the priest.

'It does me good to hope, and also to believe firmly,' Suor Maria Celeste wrote to Galileo on 3 December, 'that His Lordship the Ambassador, when he departs from Rome, will be bringing you the news of your dispatch, and also word that he personally will conduct you here in his company. I do not believe that I will live to see that day. May it please the Lord to grant me this grace, if it be for the best.'

Suor Maria Celeste's fear of not living to see Galileo's return might have stemmed from some morbid presentiment regarding her indifferent health, but more likely it arose from her frustration at the innumerable delays and rumours bearing false hope. 'I understand everyone in Florence is saying you will soon be here,' she wrote the following week, 'but until I have this from your lips, all I will believe is that your dear friends are allowing their affections and desires to give themselves voice.'

This time, however, the hearsay held true. Urban had condescended at last to change Galileo's place of imprisonment to Arcetri – not so much to commute his sentence as to make it harsher, since the ambience at Siena approached that of an exclusive salon. The pope's recommendation to the Holy Office stipulated that Galileo be limited in his social contacts henceforward, and that he refrain ever after from all teaching activities. Under these conditions, he now would be allowed to go home.

$\mathcal{M}$ OST BELOVED LORD FATHER

ONLY A MOMENT BEFORE the news of your dispatch reached me, Sire, I had taken my pen in hand to write to Her Ladyship the Ambassadress to beg her once more to intercede in this affair; for having watched it wear on so long, I feared that it might not be resolved even by the end of this year, and thus my sudden joy was as great as it was unexpected: nor are your daughters alone in our rejoicing, but all these nuns, by their grace, give signs of true happiness, just as so many of them have sympathised with me in my suffering.

We are awaiting your arrival with great longing, and we cheer ourselves to see how the weather has cleared for your journey.

Signor Geri was leaving this morning with the Court [for the annual winter session at Pisa], and I made sure to have him notified before daybreak of your return, Sire; seeing as he had already learned something of the decision, and came here last evening to tell me what he knew.

I also explained to him the reason you have not written to him, Sire, and I bemoaned the fact that he will not be here when you arrive to share in our celebration, since he is truly a perfect gentleman, honest and loyal.

I set aside the container of verdea wine, which Signor Francesco could not bring along because his litter was too overloaded. You will be able to send it to the Archbishop later, when the litter makes a return trip: the citron sweetmeat morsels I have already consigned to him. The casks for the white wine are all in order.

More I cannot say for the dearth of time, except that all of us send you our loving regards.

FROM SAN MATTEO, THE 10TH DAY OF DECEMBER 1633.
Your most affectionate daughter,

S. M. Celeste

Here she was rushing to get her last letter to her father into the hands of Signor Francesco Lupi, Suor Maria Vincenzia's brother-in-law, before his litter rumbled off to Rome by way of Siena. Although the Niccolinis could not accompany Galileo back to Arcetri as they had intended, Galileo indeed returned by the end of the week. Grand Duke Ferdinando came in person to Il Gioiello to welcome him back and stayed to visit for two hours. They spoke of life and honour, and how Galileo had preserved his against formidable odds, to become even more esteemed in his patron's eyes. If Ferdinando's fidelity to Galileo had fluttered briefly during the trial in response to Urban's threats, the future would find him a more steadfast friend.

On 17 December, Galileo wrote a formal letter of thanks to his most highly placed supporter in Rome, Francesco Cardinal Barberini:

> I have always taken special note of how affectionately Your Eminence has empathised with me in the events that befell me, and I especially recognise the value of your intercession in ultimately securing for me the grace of my being allowed to return to the quiet of my villa, precisely as I wanted to do. This and a thousand other kindnesses, all originating from your benign hand, confirm in me the wish, no less than the obligation, to always serve and revere Your Eminence, whenever it may please you to honour me with your command: not having such an order from you at the moment, I render the requisite thanks for the favour received, which I so

fervently desired; and with the most respectful love I bow to
you and kiss your robe, wishing you every happiness this
most holy Christmas.

In truth Galileo was not so much home now as under perpetual
house arrest. Later he would dateline his letters, 'From my prison
in Arcetri'. He was forbidden to receive any visitors who might
discuss scientific ideas with him. Nor could he go anywhere except
to the neighbouring convent, where the private reunion with his
daughters revealed the true emotional cost to Suor Maria Celeste
of the long, anxious separation. She had been frequently ill, he
discovered, but had paid too little attention to herself.

Galileo might have expected her to regain her stamina now in
the relief of his repatriation and the sudden respite from responsibility for his affairs. But instead she grew weaker.

'Most of all I am distressed by the news of Suor Maria Celeste,'
Niccolò Aggiunti wrote
from Pisa when Galileo
told him of her condi-
tion. 'I know the fatherly
and daughterly affection
which exists between you;
I know the lofty intellect,
and the wisdom, prudence
and goodness with which
your daughter is en-
dowed, and I know of no
one who in the same way
as she remained your
unique and gentle com-
forter in your tribu-
lations.'

For months she had
dropped all talk of entering the other life, to focus
only on having her father

Unsigned, undated portrait thought to be of
Suor Maria Celeste

return to his home and their life together. But now it seemed that both those prayers might be answered simultaneously.

In the weakened state she had described so often, Suor Maria Celeste easily succumbed to one of the many contaminants in the food or water supply. Towards the end of March 1634, she fell gravely ill with dysentery. From the moment she took sick, Galileo walked from Il Gioiello to San Matteo every day, trying to hold on to her with love and prayer. The disease cursed her with intense, unremitting abdominal pain. Her inflamed intestines evacuated fluids indiscriminately, some blood along with the vital water, until she became dehydrated. The tiny amounts of broth she could swallow would not revive her, and finally the whole balance of her body tipped against her heart. Despite the best efforts of Doctor Ronconi and Suor Luisa to save her, she died during their vigil on the second night of April.

Galileo's grief felled him. For months he sought his only solace in reading religious poems and dialogues.

'The death of Suor Maria Celeste still tears at my heart', the ambassadress Caterina Niccolini said in condolences sent from Rome on 22 April, 'like the love I bore her on account of her most virtuous nature, as well as those traits she inherited from Your Lordship, with whom I sympathise completely in this torment and in all else you have suffered.'

The archbishop of Siena apologised that he could find no words to console his friend on the loss of such a daughter, but he tried nevertheless, and he counselled Galileo to summon all his forbearance and strength for this current trial. 'I have known for a long time that she was the greatest good Your Lordship had in this world,' the archbishop wrote, 'and of such towering personal importance as to merit more than paternal love. But her having employed her spirit in the service of the next life now grants her the privilege of that singular charity, enabling her to transcend our human plight, so that she deserves to be envied rather than pitied.'

Geri Bocchineri rued the irony that Suor Maria Celeste, truly worthy of living for centuries, had followed the all too human course of dying young. 'A father who turns his tender love towards a most virtuous, most reverent daughter', Signor Geri wrote to Galileo, 'cannot deny himself the full expression of his loss at her departure; of necessity, his tears must fall. But Your Lordship can cherish the hope that a maiden so good and holy will make her way straight to the Lord God, and pray for you there before Him, and so you may reconcile yourself to that encounter, and be consoled, rather than rail against the death that has placed her in Heaven, for I believe we will need to entreat her far more than she will ever need our prayers. Always have I admired and esteemed her, and never once did I leave her presence without feeling edified, moved, contrite. Surely blessed God has already gathered her into His arms.'

As Galileo received the comfort of these words, he still suffered the effects of his physical frailties, including the aggravation of his hernia. These problems mixed with his unhappiness to produce an irregular pulse and heart palpitations.

'I feel immense sadness and melancholy,' Galileo confided to Signor Geri at the end of April, 'together with extreme inappetite; I am hateful to myself, and continually hear my beloved daughter calling to me.'

Galileo's son, Vincenzio, chose this difficult moment to make his own pilgrimage to the Casa Santa in Loreto, and from there to assume a series of law clerkships outside Florence, against his father's objections.

'I do not think it proper that Vincenzio should leave me for his travels,' the bereft father complained to Signor Geri, 'since from one hour to the other something might happen which would make his presence useful, because in addition to all this [mourning and sickness] a perpetual sleeplessness makes me afraid.'

In July, in a letter to Elia Diodati in Paris, Galileo explained Suor Maria Celeste's death in the context of his punishment and limited future.

> I stayed five months at Siena in the house of the archbishop;
> after which my prison was changed to confinement to my

own house, that little villa a mile from Florence, with strict injunctions that I was not to entertain friends, nor to allow the assembling of many at a time. Here I lived on very quietly, frequently paying visits to the neighbouring convent, where I had two daughters who were nuns and whom I loved dearly, but the eldest in particular, who was a woman of exquisite mind, singular goodness, and most tenderly attached to me. She had suffered much from ill health during my absence, but had not paid much attention to herself. At length dysentery came on, and she died after six days' illness, leaving me in deep affliction. And by a sinister coincidence, on returning home from the convent, in company with the doctor who had just told me her condition was hopeless and she would not survive the next day, as indeed came to pass, I found the Inquisitor's Vicar here, who informed me of a mandate of the Holy Office at Rome that I must desist from asking for grace or they would take me back to the actual prison of the Holy Office. From which I can infer that my present confinement is to be terminated only by that other one which is common to all, most narrow, and enduring for ever.

Into Galileo's morbid house at this juncture, his widowed sister-in-law, Anna Chiara Galilei, brought three daughters and her youngest son, Michelangelo, only to perish there with them during a brief reprise of the plague in 1634. Then Galileo, in his loneliness, invited another son of Anna Chiara's to stay with him – Alberto, Suor Maria Celeste's 'adorable little Albertino', who now worked as violinist and lutenist to the elector in Germany. The two men comforted each other for a time at Il Gioiello until Alberto went back to Munich to marry.

Now there was nothing for Galileo to do but lose himself in his work. In August he resumed active correspondence with fellow mathematicians, and in the autumn he reopened the unfinished manuscript for *Two New Sciences*.

[XXXII]

As I struggle to understand

THE EXPERIENCE OF RESUMING with Salviati, Sagredo and Simplicio – at the age of seventy – the topics that had engaged him since his first awakening as a philosopher doubly challenged Galileo. On the one hand, his ever-accumulating wisdom helped him regard certain ancient concepts in fresh ways, and this delayed his bringing the long unfinished work to closure even now. 'The treatise on motion, all new, is in order,' he wrote to an old friend in Venice, 'but my unquiet mind will not rest from mulling it over with great expenditure of time, because the latest thought to occur to me about some novelty makes me throw out much already found there.'

On the other hand, his accumulated years hampered the alacrity of his thought. 'I find how much old age lessens the vividness and speed of my thinking,' Galileo wrote to Elia Diodati while completing *Two New Sciences*, 'as I struggle to understand quite a lot of things I discovered and proved when I was younger.'

But where and how would he publish the product of all this effort? Certainly not in Rome or Florence. Shortly before Galileo returned to Arcetri, Pope Urban had issued a companion warning to the banning of the *Dialogue*, outlawing the reprinting of any of Galileo's earlier books. This action ensured that Galileo's works would gradually die out in Italy, where the Holy Office exerted its greatest influence.

'You have read my writings', Galileo complained of the prohibition against him to another correspondent in France,

> and from them you have certainly understood which was the true and real motive that caused, under the lying mask of religion, this war against me that continually restrains and undercuts me in all directions, so that neither can help come to me from outside nor can I go forth to defend myself, there having been issued an express order to all Inquisitors that they should not allow any of my works to be reprinted which had been printed many years ago or grant permission to any new work that I would print . . . a most rigorous and general order, I say, against all my works, *omnia edita et edenda* [everything published and everything I might have published in the future]; so that it is left to me only to succumb in silence under the flood of attacks, exposures, derision and insult coming from all sides.

Galileo's friend Fra Fulgenzio Micanzio, theologian to the Venetian republic, thought he could get around the pontifical warnings to see *Two New Sciences* published in the more liberal atmosphere at Venice. Fra Micanzio soon discovered in preliminary conversations with the Venetian inquisitor, however, that Galileo faced the same obstacles there as in any other Italian duchy or papal state –

that even the Credo or the Lord's Prayer might well be refused a printing licence if Galileo were the one to seek it.

There ensued a multinational effort among Galileo's supporters to find a printer somewhere who could translate and safely publish *Two New Sciences*. Geneva-born Parisian Elia Diodati hoped at first to see this happen in France, in the city of Lyons, the home of Galileo's distant relative Roberto Galilei, a businessman who facilitated all French correspondence with the Italian

GALILEVS GALILEI FLORENTINVS
ANNVM AGENS LXXVIII

Engraving of Galileo, by Francesco Zucchi

scientist. However, Galileo soon had another offer of publication help in 1635 from an Italian engineer working for the Holy Roman emperor and eager to have *Two New Sciences* printed in Germany. Grand Duke Ferdinando voluntarily lent his aid to this plan, commissioning his brother Prince Mattia, who was conveniently just leaving for Germany on a military mission, to hand-deliver sections of the contraband manuscript to Galileo's contact there. Alas, Father Christopher Scheiner, the Jesuit astronomer formerly known as 'Apelles', had returned to Germany by this point, strengthening the anti-Galileo feelings in that country and making the licensing of the new book there highly unlikely.

At the end of various intrigues, Diodati found Galileo a Dutch publisher, Louis Elzevir, who visited him at Il Gioiello in May of 1636 to settle their agreement. (Although Galileo was now technically forbidden to receive visitors, Elzevir numbered among several distinguished foreign callers, including philospher Thomas

Hobbes, who came after reading an unauthorised English translation of the *Dialogue*, and poet John Milton.)* Fra Micanzio in
Venice, who knew both parties to the publishing contract, volun-
teered to serve as conduit between Arcetri and Holland; this gave
the old theologian the pleasure of reading *Two New Sciences* in
instalments as each finished part reached him.

'I see that you took the trouble to transcribe these in your own
hand', Fra Micanzio once remarked with surprise upon receipt of
certain pages, 'and I don't see how you can stand it, for to me it
would be absolutely impossible.'

While Galileo refined the main themes, he also expanded the
content of the book to include some seemingly unrelated sections.
After all, who knew when he would ever secure another
opportunity to publish anything?

'I shall send as soon as possible this treatise on projectiles,'
Galileo promised in December 1636 while finalising Day Four of
Two New Sciences, 'along with an appendix [twenty-five pages long]
on some demonstrations of certain conclusions about the centres
of gravity of solids, found by me at the age of 22 after two years
of study of geometry, for it is good that these not be lost.'

In June of 1637, Galileo sent off the last pieces of dialogue for
Two New Sciences, which ended with Sagredo's hopeful allusion to
other discussion meetings the trio might enjoy 'in the future'.
Printing began at Leiden, Holland, that autumn, and the published
volume came out the following spring.

Safe in a Protestant country, the Dutch publisher feared no
reprisal from the Roman Inquisition. Galileo, however, remaining
vulnerable in Arcetri, claimed ignorance of the book's publication
until the ultimate moment. Even in his dedicatory note to French

---

* In 1644, in his prose polemic *Areopagitica* defending freedom of the press, Milton wrote:
'I have sat among their learned men and been counted happy to be born in such a place
of philosophic freedom as they supposed England was, while they themselves did nothing but bemoan the servile condition into which learning amongst them was brought;
that this was it which had damped the glory of Italian wits, that nothing had been there
written now these many years but flattery and fustian. There it was that I found and
visited the famous Galileo, grown old, a prisoner of the Inquisition.'

ambassador François de Noailles, he feigned surprise at how his manuscript had found its way to a foreign printing press. 'I recognise as resulting from Your Excellency's magnanimity the disposition you have been pleased to make of this work of mine,' Galileo wrote in a preface dated 6 March 1638,

John Milton visiting Galileo at Il Gioiello

notwithstanding the fact that I myself, as you know, being confused and dismayed by the ill fortune of my other works, had resolved not to put before the public any more of my labours. Yet in order that they might not remain completely buried, I was persuaded to leave a manuscript copy in some place, that it might be known at least to those who understand the subjects of which I treat. And thus having chosen, as the best and loftiest such place, to put this into Your Excellency's hands, I felt certain that you, out of your special affection for me, would take to heart the preservation of my studies and labours. Hence, during your passage through this place on your return from your Roman embassy, when I was privileged to greet you in person (as I had so often greeted you before by letters), I had occasion to present to you the copy that I then had ready of these two works. You benignly showed yourself very much pleased to have them, to be willing to keep them securely, and by sharing them in France with any friend of yours who is apt in these sciences, to show that although I remain silent, I do not therefore pass my life in entire idleness.

I was later preparing some other copies, to send to Germany, Flanders, England, Spain, and perhaps also to some place in Italy, when I was notified by the Elzevirs that they had these works of mine in press, and that I must therefore decide about the dedication and send them promptly my thoughts on that subject. From this unexpected and astonishing news, I concluded that it had been Your Excellency's wish to elevate and spread my name, by sharing various of my writings, that accounted for their having come into the hands of those printers who, being engaged in the publication of other works of mine [Letter to the Grand Duchess Cristina], wished to honour me by bringing these also to light at their handsome and elaborate press ... Now that matters have arrived at this stage, it is certainly reasonable that, in some conspicuous way, I should show myself grateful by recognising Your Excellency's generous affection. For it is you who have thought to increase my fame by having these works spread their wings freely under an open sky, when it appeared to me that my reputation must surely remain confined within narrower spaces.

Around the same time he wrote this fictitious scenario, Galileo appealed to the Holy Office for permission to seek medical treatment in Florence. Urban's brother Antonio Cardinal Sant' Onofrio, sternly denying this request via the Florentine inquisitor, ruled that Galileo had not described his illness in enough detail to be granted such an indulgence. Furthermore, the cardinal imagined, 'Galileo's return to the city would give him the opportunity of having meetings, conversations and discussions in which he might once again let his condemned opinions on the motion of the Earth come to light.'

Galileo's failing health forced him to persist in this pursuit, however, and after he submitted to a surprise medical examination demanded by the Inquisition, he won the right to repair temporarily to Vincenzio's house on the Costa San Giorgio. On 6 March Cardinal Sant' Onofrio told the inquisitor at Florence that Galileo 'may let himself be moved from the villa at Arcetri, where he now is, to his house in Florence in order to be cured of his maladies.

But I give Your Eminence orders that he must not go out into the city and not have public or secret conversations at his house.'

From the city, Galileo petitioned again, asking allowance to be carried in his chair by his family, over the few steps he could not walk in his present state, to hear mass at the neighbourhood Church of San Giorgio. In the spirit of Easter, Antonio Cardinal Sant' Onofrio then instructed the Florentine inquisitor, 'according to his own judgment to give Galileo permission to attend Mass on feast days in the nearest possible church, provided that he does not have personal contacts'.

Galileo returned to Arcetri later in the spring of 1638, before *Two New Sciences* came off the printing press at Leiden. Somehow his title for the book got changed, if not in translation, perhaps in translocation or by editorial fiat. Its title page names it:

Discourses
and
Mathematical
Demonstrations,
Concerning Two New Sciences
Pertaining to
Mechanics & Local Motions,
by Signor
Galileo Galilei Lyncean
Philosopher and Chief Mathematician to His Serene Highness
The Grand Duke of Tuscany.
With an Appendix on the centre of gravity in various Solids.

No record remains of Galileo's original title, but only his later lament over the substitution of 'a low and common title for the noble and dignified one' he had selected. Nevertheless, the book sold briskly when it appeared in June of 1638. Weeks passed after its publication before Galileo himself received even a single copy. And by the time it reached his hands, he could not read or even see it. His eyes, vulnerable to infections and strains that had pained

him much of his life, were now ruined by a combination of catar-
acts and glaucoma.

The blindness took first his right eye, in July 1637, forcing him
to abandon the addition of a fifth day to *Two New Sciences*, and
then the left the next winter. During the gloaming time when he
had only one eye with which to observe the heavens or peruse his
earlier notes and drawings, he wrote a final brief treatise on how
best to gauge the diameters of stars and the distances between
celestial bodies, and also made his last astronomical discovery,
regarding the librations, or rocking, of the Moon.

'I have discovered a very marvellous observation in the face of
the Moon,' Galileo wrote to Venetian Fra Micanzio in November
of 1637, 'in which body, though it has been looked at infinitely
many times, I do not find that any change was ever noticed, but
that the same face was always seen the same to our eyes.'

The Moon indeed keeps the same face – that of a smiling man's
eyes, nose and mouth – always turned towards the Earth. For
although the Moon rotates about its axis as it revolves around the
Earth, the time period of its rotation precisely matches the
monthly period of its revolution, keeping the far side out of sight.*
Around the fringes of the Moon's face, however, a combination
of curious effects affords occasional glimpses of parts otherwise
unseen.

'It alters its aspect', Galileo told Fra Micanzio, 'like one who
shows to our eyes his full face, head on so to speak, and then goes
changing this in all possible ways, that is, turning now a bit to the
right and then a bit to the left, or else nodding up and down, or
finally, tilting his left shoulder to right and left. All these variations
are seen in the face of the Moon, and the large and ancient spots
perceived in it make manifest and sensible what I say.'

When the total darkness descended, Galileo tried to accept the
loss of his sight gracefully, remarking how no son of Adam had
seen further than he. Still the irony overwhelmed him.

---

* It remained thus hidden until the start of the space age, when the unmanned Russian
Luna 3 spacecraft radioed the first views of the Moon's far side from lunar orbit in 1959.

'This universe,' he railed to Elia Diodati in 1638, 'which I with my astonishing observations and clear demonstrations had enlarged a hundred, nay, a thousandfold beyond the limits commonly seen by wise men of all centuries past, is now for me so diminished and reduced, it has shrunk to the meagre confines of my body.'

[XXXIII]

The memory
of the
sweetnesses

WHILE GALILEO GREW OLD and bent under house arrest in Arcetri, prohibited by inquisitors and infirmities from leaving Il Gioiello, the priest assigned to San Matteo visited him once a month, the convent records show, presumably to hear his confession and administer the sacrament of the Eucharist.

Thus enfeebled, Galileo welcomed the October 1638 arrival of the perfect live-in companion: Vincenzio Viviani, a Florentine youth of sixteen years with a remarkable aptitude for mathematics. His scholastic distinction had brought the boy to the attention of Grand Duke Ferdinando, who in turn commended him to Galileo as an assistant.

Vincenzio Viviani

Viviani wrote Galileo's letters for him, read aloud the responses, and helped Galileo reconstruct his earliest scientific investigations to clarify questions raised by correspondents. The biography of Galileo that Viviani began writing years later, in 1654, suggests a timeless period of pleasant hours shared by these two alone, when the old man would unleash his tongue to ramble and the young one would listen with all his might. It was Viviani who pursued, perpetuated and all but mythologised certain pivotal moments in the story of Galileo's life – how, for example, while still a medical student, he intuited the law of the pendulum from watching a lamp swing during mass in the Pisan cathedral,* and how he dropped cannonballs from the summit of the Leaning Tower to crowds of professors and students below.

If Galileo warmed to Viviani as to a second son, he also enjoyed the attentions of his actual son through these last years. Vincenzio, now the father of three boys (the youngest, Cosimo, was born in 1636), visited Galileo in Arcetri – and most likely Suor Arcangela as well, who lived on in silence at the convent next door. When Galileo conceived the idea for harnessing the pendulum as the regulator for a mechanical clock, he discussed the project at length

---

* Pisan tour guides to this day point out 'Galileo's lamp', though the cathedral contains at least a dozen likely candidates, and the designated attraction was installed in 1587, *after* Galileo reportedly experienced his epiphany in 1582. Regardless, all lamps swing in obedience to the same laws of physics Galileo discerned.

with his son, appealing to Vincenzio for the use of his sight and artistic skill in drawing the clockwork. The sketch completed, Vincenzio offered to build the working model himself, rather than let the idea fall into the hands of some competitor who might pirate Galileo's invention.

In a description he later wrote of his father, Vincenzio mixed memories of such times with hagiography:

> Galileo was of jovial aspect, in particular in old age, of proper and square stature, of robust and strong complexion, as such that is necessary to support the really Atlantic efforts he endured in the endless celestial observations. His eloquence and expressiveness were admirable; talking seriously he was extremely rich of sentences and deep concepts; in the pleasing discourses he did not lack wit and jokes. He was easily angered but more easily calmed. He had an extraordinary memory, so that, in addition to the many things connected to his studies, he had in mind a great quantity of poetry and in particular the better part of *Orlando Furioso*, his favourite among the poems of whose author [Ludovico Ariosto] he praised above all the Latin and Tuscan poets. His most detested vice was the lie, maybe because with the help of the mathematical science he knew the beauty of the Truth too well.

In 1641 Benedetto Castelli, through persistent petitioning of the Holy Office, obtained permission to come to Arcetri and study the motions of the Jovian moons with his old teacher, as well as to advise him spiritually – with the caveat that any discussion of the *Earth*'s motion would be grounds for excommunication.

'The falsity of the Copernican system must not on any account be doubted,' Galileo affirmed in his correspondence at this time,

> especially by us Catholics, who have the irrefragable authority of Holy Scripture interpreted by the greatest masters in theology, whose agreement renders us certain of the stability

of the Earth and the mobility of the Sun around it. The conjectures of Copernicus and his followers offered to the contrary are all removed by that most sound argument, taken from the omnipotence of God. He being able to do in many, or rather in infinite ways, that which to our view and observation seems to be done in one particular way, we must not pretend to hamper God's hand and tenaciously maintain that in which we may be mistaken. And just as I deem inadequate the Copernican observations and conjectures, so I judge equally, and more, fallacious and erroneous those of Ptolemy, Aristotle and their followers, when without going beyond the bounds of human reasoning their inconclusiveness can be very easily discovered.

Upon Castelli's return to Rome, he resumed his efforts to see Galileo's sentence of house arrest commuted, though these proved unsuccessful. Castelli continued to say mass for Galileo every morning (until his own death in 1643), and the two close friends kept in touch on matters of mutual interest. Concluding a letter to Castelli on the hydraulics of fountains and rivers, Galileo expressed gratitude for the solace of his companionship over a lifetime: 'Bereft of my powers by my great age and even more by my unfortunate blindness and the failure of my memory and other senses, I spend my fruitless days which are so long because of my continuous inactivity and yet so brief compared with all

Painting of Galileo at age seventy-one, by Justus Sustermans

the months and years which have passed; and I am left with no other comfort than the memory of the sweetnesses of former friendships, of which so few are left, although one more undeserved than all the others remains: that of corresponding in love with you.'

Still Galileo's imagination would not rest but returned to the longitude problem, to the books of Euclid, whose ancient definitions of ratios Galileo redefined at this time, as well as to any number of other ideas he may have entertained but neither recorded nor reported to anyone.

'I have in mind a great many miscellaneous problems and questions,' he wrote to a fellow philosopher in Savona, 'partly quite new and partly different from or contrary to those commonly received, of which I could make a book more curious than the others written by me; but my condition, besides blindness on top of other serious indispositions and a decrepit age of 75 years, will not permit me to occupy myself in study. I shall therefore remain silent, and so pass what remains to me of my laborious life, satisfying myself in the pleasure I shall feel from the discoveries of other pilgrim minds.'

One such roving intellect, the self-pronounced 'Galileist' Evangelista Torricelli, who had been one of Castelli's most precocious students in Rome, sent Galileo a letter and a manuscript for comment. Impressed, Galileo invited him to Arcetri. 'I hope to enjoy your company for some few days before my life, now near an end, is finished,' Galileo wrote to Torricelli in September 1641, 'as also to discuss with you some relics of my thoughts on mathematics and physics and to have your aid in polishing them, that they may be left less messy to be seen with other things of mine.'

Torricelli moved into Galileo's house in October. The very next month, however, Galileo took to his canopy bed with a fever and pain in his kidneys that presently proved fatal. As he lay dying for more than two months, Galileo had time to dictate to Torricelli some thoughts in dialogue on mathematical ratios. It was the beginning of a new book – another intellectual adventure for Salviati, Sagredo and Simplicio – but it was too late. On the evening of

8 January 1642, Galileo died in the quiet company of Torricelli, Viviani and his own Vincenzio.

'Today the news has come of the loss of Signor Galilei,' wrote Lucas Holste, Francesco Cardinal Barberini's Vatican librarian, 'which touches not just Florence but the whole world, and our whole century which from this divine man has received more splendour than from almost all the other ordinary philosophers. Now, envy ceasing, the sublimity of that intellect will begin to be known which will serve all posterity as guide in the search for truth.'

Though Holste's eulogy would eventually prove prophetic, the lingering effects of Galileo's trial and condemnation subdued the immediate reaction to his death.

Grand Duke Ferdinando, dubbing Galileo 'the greatest light of our time', buried him in the novices' chapel in the Franciscan church of Santa Croce. Ferdinando had hoped to honour the wishes expressed in Galileo's will, that he be buried next to his father and other relatives in the church's main basilica, which was padded with private chapels bearing the tombs and coats of arms of many of Florence's finest families. Beyond this, the grand duke had proposed a public funeral oration and the erection of a marble mausoleum, but Pope Urban called down from Rome to check all these attempts. Any such fuss over the dead body of Galileo, Urban announced with an anger that he took to his own grave, would be deemed an offence against papal authority.*

Ferdinando submitted to the pope's decree. He dropped the idea of the monument, and, eschewing the burial site inside the basilica, he had Galileo's body interred in a closet-sized room off the chapel, under the campanile, where he naturally assumed it would stay. But the nineteen-year-old Viviani, with a dedication

---

* When Urban died on 29 July 1644, the people of Rome expressed their resentment of his last expensive war, begun in Castro in 1641, by demolishing a statue of him in the courtyard of the Collegio Romano. 'The pope died at quarter past eleven,' a diarist noted, 'and by noon the statue was no more.' The Thirty Years' War, which had raged on despite Urban's interventions, finally ended in 1648 with the Peace of Westphalia.

born of devotion, committed himself to moving and correctly commemorating his mentor's mortal remains, some day, somehow.

Though Vincenzio Galilei was heir to his father's financial estate – except for an annuity of thirty-five *scudi* set aside for the lifelong support of Suor Arcangela – Vincenzio Viviani inherited the master's mathematics. Having studied as Galileo's pupil, Viviani succeeded Torricelli, in 1647, in his teacher's former post as mathematician to Grand Duke Ferdinando. And having served as Galileo's amanuensis, Viviani later gathered all of his papers to publish, in 1656, the first edition of the collected works of Galileo – without the *Dialogue*, of course.

Ferdinando de' Medici and his brother Prince Leopold inducted Viviani as a charter member of their scientific society, the Accademia del Cimento, which began holding regular meetings at the Pitti Palace in 1657. That honour, coupled with the lustre of his own published books on mathematics, helped spread Viviani's fame, so that King Louis XIV appointed him one of only eight foreign members of the newly founded French Académie Royale des Sciences in 1666.

All this time Viviani never slackened his resolve to resuscitate the plans for the monument and public recognition of Galileo. He received scant assistance from Galileo's family, understandably, for Vincenzio died of a fever in 1649, and Suor Arcangela, though she survived another ten years after her brother's death, could not have been any help.

Viviani found a sculptor to create Galileo's death mask, and from it a bust. Later he commissioned a different sculptor to fashion a second bust in bronze, then a marble version from yet another artist whom he charged to design the still forbidden tomb. Viviani assumed all responsibility for the execution and expense of these efforts, trying simultaneously to convince influential persons of their importance. He lost Ferdinando's support in 1670, when the grand duke's five-decade reign ended in an agony of dropsy, apoplexy and the painful treatments of his doctors. Viviani turned hopefully to the new heir to the throne, Ferdinando's son, the gluttonous, sanctimonious Cosimo III, but he proved to be an ineffectual leader with no interest in science, who taxed the Florentines unfairly as he squandered what remained of the Medici fortune.* In Rome, Urban's immediate successors – Innocent X for ten years, followed by Alexander VII for twelve – focused their funerary attention on their own lavish tombs.

In September 1674, in frustration, Viviani mounted a marble plaque and a plaster bust on the wall of the tiny room off the novices' chapel where Galileo still lay in a virtually unmarked grave. He followed this action with a unique public yet private tribute, by renovating the façade of his own new house to accommodate a bust of Galileo over the arched front doorway, with eulogies to him inscribed in enormous stone scrolls on either side.

When Viviani died childless in 1703 at the age of eighty-one, he left all his worldly goods – including the solemn responsibility for relocating Galileo's grave – in the hands of a nephew who never accomplished the plan in the remaining three decades of his own life. Viviani's property, along with the duty it implied, passed eventually by family inheritance to a senator of Florence, Giovanni Battista Clemente de Nelli, who enjoyed the triumphant satisfaction of finally seeing the Galileo tomb project through to completion in 1737, thanks in part to the accession of Lorenzo Corsini as Pope Clement XII. With a Florentine pontiff once more ensconced on the throne of Saint Peter, Galileo would finally claim his proper due.

The mausoleum had first taken form in Ferdinando's and Viviani's envisionings as a match for the ornate monument to Michelangelo Buonarroti, another fellow Tuscan, near the entrance to Santa Croce. There against the south wall of the church, Michelangelo's highly perched bust oversees the marble forms of the muses of Painting, Sculpture, and Architecture, who sit around his coffin in mourning. Not only did Ferdinando and Viviani view Galileo's genius as the scientific counterpoint to Michelangelo's art,

---

* Within one more generation, the great chain of government that had dominated Tuscany's political structure since the fourteenth century would die out with the last Medici grand duke, Gian Gastone, in 1737.

Galileo offering his telescope to the Muses

but Viviani also promulgated the belief that Michelangelo's spirit had leaped like an inspiration from his aged, failing body to the infant Galileo in the brief span of hours separating the former's death from the latter's birth.

The original design for Galileo's monument called for three female forms to attend him – the muses of Astronomy, Geometry and Philosophy, who would stand symmetrically opposed to

Michelangelo's three seated muses of the arts. In the tomb's eventual execution, however, only two such statues, Astronomy and
Geometry, emerged to flank the bust of Galileo, who holds in his
right hand a telescope, while his left rests on a globe and an assortment of books. Philosophy was omitted, either on instructions
from the Holy Office or out of fear that her presence would recall
unfortunate memories of Galileo's condemnation. And yet there
is a third female figure incorporated into the tomb – in such a way
that she remains invisible to even the most careful observer.

On the evening of 12 March 1737, after permission had been
granted to move Galileo's body from its first interment site to the
marble sarcophagus of the nearly completed monument, a distinguished congregation – part ecclesiastic, mostly civic – assembled
discreetly with torches and candles in the Church of Santa Croce.
Their task that long night involved elements of demolition work,
religious ceremony, mortuary identification and hero veneration
as they exhumed the body of Galileo. This occasion also saw the
ritual removal of a single vertebra from the venerated scientist,
along with three fingers of his right hand and a tooth – and surely
would have included the preservation of his brain as well, if by
some miracle that organ had still survived.

The closet-sized room under the campanile, where Galileo had
lain buried for ninety-five years, now contained two brick biers:
one belonging to Galileo and the other to his disciple, Vincenzio
Viviani, who had asked in his will to share the master's grave.

The few men who could fit inside the tiny room broke open
the more recent brickwork, laid at the time of Viviani's death in
1703, and extracted a wood coffin. According to an eyewitness
report filed by a notary, they carried this into the novices' chapel,
where everyone could watch as the lid was lifted to reveal a lead
plate identifying the corpse as Viviani's. Several sculptors and
scientists in the party covered the bier with a black cloth,
and lifting the draped coffin to their shoulders, they bore it
through the long passageway from the chapel and across the
cavernous basilica. Their chanted prayers for the dead reverberated off the wooden columns, which towered over the unattended

procession, and the stone walls that had been frescoed by Giotto to trace the life of Saint Francis.

The assembly placed the body at the new site, then returned to the little chapel and set about repeating the procedure – smashing the older brick container under the 1674 memorial Viviani had mounted for Galileo, and pulling out another wooden coffin. This one had apparently been damaged over time, its lid bashed in and littered with broken pieces of plaster. As the men dragged the bier

Monumento eretto a Galileo Galilei in Firenze nella Chiesa di S. Croce. Opera delli Scultori Foggini.

Galileo's tomb in Santa Croce

from the bricks, they were startled to discover another almost identical wooden box lying directly beneath it. Galileo's grave contained two coffins, two skeletons, and no lead nameplate on either one of them.

Panic no doubt gripped several hearts at the prospect of being unable to decide which body deserved to be deposited in the new monument. But when the grand duke's chief physician, accompanied by several professors of anatomy, stepped forward to examine the evidence, they accomplished their identification with reassuring ease. Only one of the skeletons could possibly have belonged to Galileo – the top one, because its bones were those of an old man, with the detached mandible containing only four teeth. The skeleton in the lower coffin, the experts all concurred, was unmistakably female. Although the woman had lain dead for at least as long a time as the man, if not longer, she had died at a much younger age.

The congregation divided itself solemnly in half, each group walking Galileo's body part-way through the basilica, so that as many participants as possible could share the honour of being his pall-bearers. Then they carried the woman to the mausoleum, too, and they laid her in the sepulchre beside her father.

Once the shock of the discovery had dissipated into the silence of the great empty church, those attendants who remembered Viviani could unfurl the mystery for themselves. The disciple, driven to despair by his failure to pay the tribute he felt he owed his mentor, had given Galileo something dearer than bronze or marble to distinguish his grave.

Even now, no inscription on Galileo's much-visited tomb in Santa Croce announces the presence of Suor Maria Celeste.

But still she is there.

In Galileo's Time

1543   Nicolaus Copernicus (1473–1543) publishes *De revolutionibus*, and Andreas Vesalius (1514–64), *On the Fabric of the Human Body*.

1545   Council of Trent convenes under Pope Paul III; first ten sessions last two years.

1551   Collegio Romano, or Pontifical Gregorian University, founded by Jesuits in Rome. Council of Trent reconvenes.

1559   First worldwide Index of Prohibited Books promulgated by the Roman Inquisition.

1562–6   Third convention and final sessions of the Council of Trent.

1564   Galileo is born in Pisa, 15 February. Michelangelo Buonarroti dies in Florence, 18 February. William Shakespeare is born in England, 23 April.

1569   Cosimo I, duke of Florence, named grand duke of Tuscany by authority of Pope Pius V.

1572   Tycho Brahe (1546–1601) of Denmark observes a nova and concludes that changes could occur in the heavens.

1577   Studies of comets by Tycho convince him the heavens could not consist of solid spheres, though he rejects the Copernican system.

1581   Galileo enrols at University of Pisa.

1582   Gregorian calendar replaces Julian in Catholic Europe.

1585   Galileo abandons studies at Pisa without a university degree.

1587   Ferdinando I becomes grand duke of Tuscany when his elder brother, Francesco, dies of malaria.

1589   Galileo begins teaching at Pisa; develops a rudimentary thermometer; begins to study falling bodies.

1591   Vincenzio Galilei (father) dies.

1592   Galileo begins teaching at the University of Padua.

1600   Giordano Bruno burns at the stake in Rome. Virginia Galilei (daughter) is born in Padua.

1601   Livia Galilei (daughter) is born in Padua. Tycho Brahe dies.

1603   Prince Federico Cesi founds Lyncean Academy in Rome.

1604   New star appears in the heavens, generating debate and three public lectures by Galileo.

1605   Prince Cosimo de' Medici takes instruction from Galileo.

1606   Galileo publishes treatise on geometric and military compass; Vincenzio Galilei (son) is born in Padua.

1607   Baldessar Capra publishes pirated Latin edition of Galileo's instructions for geometric and military compass.

1608   Hans Lippershey invents a refracting telescope in Holland. Prince Cosimo marries Maria Maddalena, archduchess of Austria.

1609   Grand Duke Ferdinando I dies; Cosimo II succeeds him. Galileo improves telescope, observes and measures mountains on the Moon. Johannes Kepler (1571–1630) publishes first two laws of planetary motion.

1610   Galileo discovers the moons of Jupiter. *The Starry Messenger* is published. Galileo is appointed chief mathematician and philosopher to the grand duke of Tuscany, Cosimo II.

1611   Galileo visits Rome, elected to membership in the Lyncean Academy.

1612   *Bodies That Stay Atop Water or Move Within It* is published in Florence.

1613   Prince Cesi publishes Galileo's *Sunspot Letters*; Virginia and Livia Galilei (daughters) enter the Convent of San Matteo in Arcetri.

1614   Virginia and Livia Galilei assume religious habit.

1616   Galileo writes his 'Theory on the Tides'. Edict issued in Rome against Copernican doctrine. Virginia Galilei professes her vows as Suor Maria Celeste. Shakespeare and Miguel de Cervantes die.

1617   Livia Galilei professes vows as Suor Arcangela.

1618   Three comets appear, generating interest and debate; Jesuit father Orazio Grassi lectures on comets at Collegio Romano; Thirty Years' War begins.

1619   Grassi's account of the comets is published anonymously; Mario Guiducci delivers *Discourse on the Comets*, provoking pseudonymous *Astronomical and Philosophical Balance*. Kepler publishes third law of planetary motion. Galileo's mistress, Marina Gamba, dies. Vincenzio Galilei (son) is legitimised.

1623   Galileo's sister Virginia dies. Maffeo Cardinal Barberini becomes Pope Urban VIII. Galileo dedicates *The Assayer* to him.

1624   Galileo travels to Rome for papal audience.

1628   William Harvey (1578–1657) in England describes the circulation of the blood.

1629   Bubonic plague enters northern Italy from Germany.

1630   Galileo visits Rome to obtain printing licence for his *Dialogue*. Prince Cesi dies. Bubonic plague strikes Florence.

1631   Michelangelo Galilei (brother) dies of plague in Germany.

1632   Galileo publishes *Dialogue on the Two Chief World Systems: Ptolemaic and Copernican*.

1633   Galileo stands trial for heresy by the Holy Office of the Inquisition; *Dialogue* is prohibited.

1634   Suor Maria Celeste Galilei dies in Arcetri on 2 April.

1636   *Letter to the Grand Duchess Cristina* is published in Holland, in Latin and Italian.

1637   Galileo discovers lunar libration, loses his eyesight.

1638   Louis Elzevir publishes Galileo's *Two New Sciences* in Leiden, Holland.

1641   Vincenzio Galilei draws his father's design for a pendulum clock.

1642   Galileo dies in Arcetri, 8 January. Isaac Newton is born in England, 25 December.

1643   Galileo's student Evangelista Torricelli (1608–47) invents mercury barometer.

1644   Pope Urban VIII dies.

1648   Thirty Years' War ends.

1649   Vincenzio Galilei (son) dies in Florence, 15 May.

1654   Grand Duke Ferdinando II improves on Galileo's thermometer by closing the glass tube to keep air out.

1655–6   Christiaan Huygens (1629–95) improves telescope, discovers largest of Saturn's moons, sees Saturn's 'companions' as a ring, patents pendulum clock.

1659   Suor Arcangela dies at San Matteo, 14 June.

1665   Jean-Dominique Cassini (1625–1712) discovers and times the rotation of Jupiter and Mars.

1669   Sestilia Bocchineri Galilei dies.

1670   Grand Duke Ferdinando II dies, succeeded by his only surviving son, Cosimo III.

1676   Ole Roemer (1644–1710) uses eclipses of Jupiter's moons to determine the speed of light; Cassini discovers gap in Saturn's rings.

1687   Newton's laws of motion and universal gravitation are published in his *Principia*.

1705   Edmond Halley (1656–1742) studies comets, realises they orbit the Sun, predicts return of a comet later named in his honour.

1714   Daniel Fahrenheit (1686–1736) develops mercury thermometer with accurate scale for scientific purposes.

1718   Halley observes that even the fixed stars move with almost imperceptible 'proper motion' over long periods of time.

1728   English astronomer James Bradley (1693–1762) provides first evidence for the Earth's motion through space based on the aberration of starlight.

1755   Immanuel Kant (1724–1804) discerns the true shape of the Milky Way, identifies the Andromeda nebula as a separate galaxy.

1758   'Halley's comet' returns.

1761   Mikhail Vasilyevich Lomonosov (1711–65) realises Venus has an atmosphere.

1771   Comet hunter Charles Messier (1730–1817) identifies a list of non-cometary objects, many of which later prove to be distant galaxies.

1781 William Herschel (1738–1822) discovers the planet Uranus.

1810 Napoleon Bonaparte, having conquered the Papal States, transfers the Roman archives, including those of the Holy Office with all records of Galileo's trial, to Paris.

1822 Holy Office permits publication of books that teach Earth's motion.

1835 Galileo's *Dialogue* is dropped from Index of Prohibited Books.

1838 Stellar parallax, and with it the distance to the stars, is detected independently by astronomers working in South Africa, Russia and Germany; Friedrich Wilhelm Bessel (1784–1846) publishes the first account of this phenomenon, for the star 61 Cygni.

1843 Galileo's trial documents are returned to Italy.

1846 Neptune and its largest moon are discovered by predictions and observations of astronomers working in several countries.

1851 Jean-Bernard-León Foucault (1819–68) in Paris demonstrates the rotation of the Earth by means of a two-hundred-foot pendulum.

1861 Kingdom of Italy proclaimed, uniting most states and duchies.

1862 French chemist Louis Pasteur (1822–95) publishes germ theory of disease.

1877 Asaph Hall (1829–1907) discovers the moons of Mars.

1890–1910 Complete works, *Le Opere di Galileo Galilei*, are edited and published in Florence by Antonio Fàvaro.

1892 University of Pisa awards Galileo an honorary degree – 250 years after his death.

1893 *Providentissimus Deus* of Pope Leo XIII cites Saint Augustine, taking the same position Galileo did in his *Letter to the Grand Duchess Cristina*, to show that the Bible did not aim to teach science.

1894 Pasteur's student Alexandre Yersin (1863–1943) discovers bubonic plague bacillus and prepares serum to combat it.

1905 Albert Einstein (1879–1955) publishes his special theory of relativity, establishing the speed of light as an absolute limit.

1908   George Ellery Hale (1868–1938) discerns the magnetic nature of sunspots.

1917   Willem de Sitter (1872–1934) intuits the expansion of the universe from Einstein's equations.

1929   American astronomer Edwin Hubble (1889–1953) finds evidence for expanding universe.

1930   Roberto Cardinal Bellarmino is canonised as Saint Robert Bellarmine by Pope Pius XI.

1935   Pope Pius XI inaugurates Vatican Observatory and Astrophysical Laboratory at Castel Gandolfo.

1950   *Humani generis* of Pope Pius XII discusses the treatment of unproven scientific theories that may relate to Scripture; reaches same conclusion as Galileo's *Letter to the Grand Duchess Cristina*.

1959   Unmanned Russian *Luna 3* spacecraft radios first views of the Moon's far side from lunar orbit.

1966   Index of Prohibited Books is abolished following the Second Vatican Council.

1969   American astronauts Neil Armstrong and Buzz Aldrin walk on the Moon.

1971   *Apollo 15* commander David R. Scott drops a falcon feather and a hammer on the lunar surface; when they fall together he says, 'This proves that Mr Galileo was correct.'

1979   Pope John Paul II calls for theologians, scholars and historians to re-examine Galileo's case.

1982   Pope John Paul II establishes Galileo Commission with four formal study groups to reinvestigate the Galileo affair.

1986   Halley's comet returns, observed by a waiting armada of spacecraft.

1989   National Aeronautics and Space Administration launches *Galileo* spacecraft to study the moons of Jupiter at close range.

1992   Pope John Paul II publicly endorses Galileo's philosophy, noting how 'intelligibility, attested to by the marvellous discoveries of science and technology, leads us, in the last analysis, to that transcendent and primordial thought imprinted on all things'.

1995   *Galileo* reaches Jupiter.

1999   *Galileo*'s successful reconnaissance of the Medicean stars, now better known as the Galilean satellites of Jupiter, continues to enlighten astronomers everywhere.

Florentine Weights, Measures, Currency

WEIGHT

libbra = 12 *oncie* = 0.75 pound = 0.3 kilogram (plural is *libbre*)

MEASURE

braccio = about 23 inches or 58 centimetres (plural is *braccia*)

CURRENCY

florin = 3.54 grammes of gold

scudo = 7 *lire*

piastra = 22.42 grammes of silver = about 5 *lire*

lira (silver coin) = 12 *crazie* = 20 *soldi* (4 *lire* could feed one person for a week)

giulio (silver coin) = slightly more than half a *lira*

carlino = 0.01 *scudo*

BIBLIOGRAPHY

Allan-Olney, Mary. *The Private Life of Galileo, Compiled Principally from His Correspondence and That of His Eldest Daughter, Sister Maria Celeste*. London: Macmillan, 1870.

Arano, Luisa Cogliati. *The Medieval Health Handbook*. New York: George Braziller, 1976, 1996.

Arduini, Carlo. *La Primogenita di Galileo Galilei rivelata dalla sue lettere*. Florence: Felice LeMonnier, 1864.

Asimov, Isaac. *Asimov's Biographical Encyclopedia of Science and Technology*. New York: Doubleday, 1972.

———. *Asimov's Chronology of Science and Discovery*. New York: Harper-Collins, 1994.

Bajard, Sophie, and Raffaello Bencini. *Villas and Gardens of Tuscany*. Paris: Terrail, 1993.

Beatty, J. Kelly, and Andrew Chaikin, eds. *The New Solar System*. 3d edn. Cambridge, Mass.: Sky Publishing, 1990.

Bedini, Silvio A. *The Pulse of Time: Galileo Galilei, the Determination of Longitude, and the Pendulum Clock*. Florence: Bibliotecca di Nuncius, 1991.

Bertola, Francesco. *Da Galileo alle Stelle*. Padua: Biblos, 1992.

Biagioli, Mario. *Galileo, Courtier*. Chicago: University of Chicago Press, 1993.

Blackwell, Richard J. *Galileo, Bellarmine, and the Bible*. Notre Dame, Ind.: University of Notre Dame Press, 1991.

Boeser, Knut, ed. *The Elixirs of Nostradamus*. London: Moyer Bell, 1996.

Bologna, Gianfranco. *Simon and Schuster's Birds of the World*. Edited by John Bull. New York: Fireside, 1978.

Bornstein, Daniel, and Roberto Rusconi, eds. *Women and Religion in Medieval and Renaissance Italy*. Translated by Margery J. Schneider. Chicago: University of Chicago Press, 1996.

Brodrick, James, SJ. *Galileo: The Man, His Work, His Misfortunes*. London: Catholic Book Club, 1964.

Brucker, Gene. *Florence: The Golden Age, 1138–1737.* Berkeley: University of California Press, 1998.

Bruno, Giordano. *The Ash Wednesday Supper / La Cena de le Ceneri.* Translated by Stanley L. Jaki. Paris: Mouton, 1975.

Bunson, Matthew. *The Pope Encyclopedia.* New York: Crown, 1995.

Calvi, Giulia. *Histories of a Plague Year: The Social and the Imaginary in Baroque Florence.* Translated by Dario Biocca and Bryant T. Ragan, Jr. Berkeley: University of California Press, 1989.

Chaikin, Andrew. *A Man on the Moon.* New York: Viking, 1994.

Cipolla, Carlo M. *Clocks and Culture, 1300–1700.* New York: Walker, 1967.

——. *Cristofano and the Plague: A Study in the History of Public Health in the Age of Galileo.* London: Collins, 1973.

——. *Faith, Reason, and the Plague in Seventeenth-Century Tuscany.* London: Harvester, 1979; New York: Norton, 1981.

——. *Fighting the Plague in Seventeenth-Century Italy.* Madison: University of Wisconsin Press, 1981.

——. *Public Health and the Medical Profession in the Renaissance.* Cambridge: Cambridge University Press, 1976.

Clare (Saint). *The Rule and Testament of Saint Clare.* Translated by Mother Mary Francis. Chicago: Franciscan Herald Press, 1987.

Cleugh, James. *The Medici: a Tale of Fifteen Generations.* New York: Doubleday, 1975.

Cohen, I. Bernard. *The Birth of a New Physics.* New York: Norton, 1985.

——. 'What Galileo Saw: The Experience of Looking Through a Telescope.' In *Homage to Galileo.* Edited by P. Mazzoldi, 445–72. Padua: Cleup, 1992.

Colette (Saint). *The Testament of Saint Colette.* Translated by Mother Mary Francis. Chicago: Franciscan Herald Press, 1987.

Coyne, G. V., M. Heller and J. Zycinski, eds. *The Galileo Affair: A Meeting of Faith and Science.* Vatican City State: Specola Vaticana, 1985.

Culbertson, Judi, and Tom Randall. *Permanent Italians: An Illustrated, Biographical Guide to the Cemeteries of Italy.* New York: Walker, 1996.

De Harsányi, Zsolt. *The Star-Gazer.* Translated by Paul Tabor. New York: Putnam, 1939.

Delaney, John. *Dictionary of Saints.* New York: Doubleday, 1980.

De Santillana, Giorgio. *The Crime of Galileo.* Chicago: University of Chicago Press, 1955.

Desiato, Luca. *Galileo Mio Padre.* Milan: Arnoldo Mondadori, 1983.

Dibner, Bern, and Stillman Drake. *A Letter from Galileo Galilei.* Norwalk: Burndy Library, 1967.

Di Canzio, Albert. *Galileo: His Science and His Significance for the Future of Man.* Portsmouth: Adasi, 1996.

DiCrollolanza, Goffredo. *Enciclopedia Araldico-Cavalleresca.* Bologna: Arnaldo Forni Editore, 1980.

Dohrn-van Rossum, Gerhard. *History of the Hour.* Translated by Thomas Dunlap. Chicago: University of Chicago Press, 1996.

Drake, Stillman. 'The Accademia dei Lincei.' *Science,* 11 March 1966, 1194–1200.

——. *Cause, Experiment, and Science.* Chicago: University of Chicago Press, 1981.

——. *Discoveries and Opinions of Galileo.* New York: Anchor, 1957.

——. *Galileo.* Oxford: Oxford University Press, 1980, reissued 1996.

——. *Galileo at Work: His Scientific Biography.* Chicago: University of Chicago Press, 1978.

——. *Galileo Studies: Personality, Tradition, and Revolution.* Ann Arbor: University of Michigan Press, 1970.

——. *History of Free Fall: Aristotle to Galileo.* Toronto: Wall and Thompson, 1989.

——. *Telescopes, Tides, and Tactics.* Chicago: University of Chicago Press, 1983.

Elliot, James, and Richard Kerr. *Rings: Discoveries from Galileo to Voyager.* Cambridge, Mass.: MIT Press, 1984.

Fahie, J. J. *Memorials of Galileo Galilei, 1564–1642.* London: Courier, 1929.

Fantoli, Annibale. *Galileo: For Copernicanism and for the Church.* Translated by George V. Coyne, SJ. Notre Dame, Ind.: University of Notre Dame Press, 1994.

Fàvaro, Antonio. *Galileo Galilei e Suor Maria Celeste.* Florence: Barbèra, 1891.

Feldhay, Rivka. *Galileo and the Church: Political Inquisition or Critical Dialogue?* Cambridge: Cambridge University Press, 1995.

Fermi, Laura, and Gilberto Bernardini. *Galileo and the Scientific Revolution*. New York: Basic Books, 1961.

Ferrari, Giovanna. 'Public Anatomy Lessons and the Carnival: The Anatomy Theatre of Bologna.' *Past and Present*, no. 117, 50–106.

Ferris, Timothy. *Coming of Age in the Milky Way*. New York: Morrow, 1988.

Finocchiaro, Maurice A. *The Galileo Affair: A Documentary History*. Berkeley: University of California Press, 1989.

———. *Galileo on the World Systems*. Berkeley: University of California Press, 1997.

Galilei, Celeste. *Lettere al Padre*. Edited by Giovanni Ansaldo (1927). Genoa: Blengino, 1992.

Galilei, Galileo. *The Assayer*. In *The Controversy on the Comets of 1618*, by Galileo Galilei, Horatio Grassi, Mario Guiducci and Johannes Kepler. Translated by Stillman Drake and C. D. O'Malley. Philadelphia: University of Pennsylvania Press, 1960.

———. *Dialogo di Galileo Galilei Linceo*. Florence: Gio.: Batista Landini, 1632.

———. *Dialogue Concerning the Two Chief World Systems*. Translated by Stillman Drake. Berkeley: University of California Press, 1967.

———. *Dialogues Concerning Two New Sciences*. Translated by Henry Crew and Alfonso de Salvio. New York: Macmillan, 1914; New York: Dover, 1954.

———. *Istoria e Dimostrazioni intorno alle Macchie Solari*. Rome: Appresso Giacomo Mascardi, 1613.

———. *Letter to Grand Duchess Cristina*. Translated by Stillman Drake. In *Discoveries and Opinions of Galileo*. New York: Anchor, 1957.

———. *Letters on Sunspots*. Translated by Stillman Drake. In *Discoveries and Opinions of Galileo*. New York: Anchor, 1957.

———. *Operations of the Geometric and Military Compass*. Translated by Stillman Drake. Washington: Smithsonian Institution Press, 1978.

———. *Opere*. 20 vols. Edited by Antonio Fàvaro. Florence: Barbèra, 1890–1909.

———. *Prose Scelte*. Collected and annotated by Professor Augusto Conti. Florence: Barbèra, 1910.

———. *Sidereus Nuncius, or The Sidereal Messenger*. Translated by Albert Van Helden. Chicago: University of Chicago Press, 1989.

———. *Two New Sciences, Including Centers of Gravity and Force of Percussion*. Translated by Stillman Drake. 2d edn. Toronto: Wall and Thompson, 1989.

Galilei, Suor Maria Celeste. *Lettere al Padre*. Edited by Giuliana Morandini. Torino: La Rosa, 1983.

Galluzzi, Paolo. 'I Sepolcri de Galileo: Le Spoglie "Vive" di un Eroe della Scienza.' In *Il Pantheon di Santa Croce a Firenze*. Edited by Luciano Berti, 145–82. Florence: Cassa di Risparmio, 1993.

Gingerich, Owen. *The Eye of Heaven: Ptolemy, Copernicus, Kepler*. New York: American Institute of Physics, 1993.

———. *The Great Copernicus Chase and Other Adventures in Astronomical History*. Cambridge, Mass.: Sky Publishing, 1992.

Godoli, Antonio, and Paolo Paoli. 'L'Ultima Dimora di Galileo: La Villa "Il Gioiello" ad Arcetri.' In *Annali dell'Istituto e Museo di Storia della Scienza*. Florence: Giunti-Barbèra, 1979.

Goodwin, Richard N. *The Hinge of the World*. New York: Farrar, Straus and Giroux, 1998.

Gribbin, John, and Mary Gribbin. *Galileo in 90 Minutes*. London: Constable, 1997.

Haggard, Howard W., MD. *Devils, Drugs, and Doctors*. New York: Blue Ribbon, 1929.

Hale, J. R. *Florence and the Medici: The Pattern of Control*. London: Thames and Hudson, 1977.

Hibbert, Christopher. *The Rise and Fall of the House of Medici*. London: Allen Lane, 1974; Penguin, 1979.

Jameson, Anna. *Legends of the Monastic Orders*. Boston: Houghton Mifflin, ca. 1840.

Kearney, Hugh. *Science and Change, 1500–1700*. New York: McGraw-Hill, 1971.

Kent, Countess of. *A Choice Manual, or, Rare and Select Secrets in Physick and Chirurgery*. London: Henry Morflock, 1682.

Kline, Morris. *Mathematical Thought from Ancient to Modern Times*. New York: Oxford University Press, 1972.

Knedler, John Warren Jr, ed. *Masterworks of Science*. Garden City: Doubleday, 1947.

Lacroix, Paul. *Science and Literature in the Middle Ages and the Renaissance*. New York: Frederick Ungar, 1878, 1964.

Landes, David S. *Revolution in Time: Clocks and the Making of the Modern World*. Cambridge, Mass.: Harvard University Press, 1983.

Langford, Jerome J. *Galileo, Science, and the Church*. Ann Arbor: University of Michigan Press, 1966; 3rd edn. 1992.

Lewis, Richard S. *The Voyages of Apollo*. New York: Quadrangle / New York Times, 1974.

McBrien, Richard P. *Catholicism*. New York: HarperCollins, 1994.

McEvedy, Colin. 'The Bubonic Plague.' *Scientific American*, February 1988, 118–23.

MacLachlan, James. *Galileo Galilei: First Physicist*. New York: Oxford University Press, 1997.

McMullin, Ernan, ed. *Galileo, Man of Science*. New York: Basic Books, 1967.

McNamara, Jo Ann Kay. *Sisters in Arms: Catholic Nuns Through Two Millennia*. Cambridge, Mass.: Harvard University Press, 1996.

Machamer, Peter, ed. *The Cambridge Companion to Galileo*. Cambridge: Cambridge University Press, 1998.

Mary Francis PCC. *Forth and Abroad*. San Francisco: Ignatius, 1997.

———. *A Right to Be Merry*. Chicago: Franciscan Herald Press, 1973.

———. *Strange Gods Before Me*. Chicago: Franciscan Herald Press, 1976.

Matter, E. Ann, and John Coakley, eds. *Creative Women in Medieval and Early Modern Italy: A Religious and Artistic Renaissance*. Philadelphia: University of Pennsylvania Press, 1994.

Micheletti, Emma. *Le Donne dei Medici*. Florence: Sansoni, 1983.

Montanari, Geminiano. *Copia di Lettera Scritta all'Illustrissimo Signore Antonio Magliabechi, Bibliotecario del Serenissimo Gran Duca di Toscana, Intorno Alla Nuova Cometa apparsa quest'anno 1682, sotto i piedi dell'Orsa Maggiore*. Padua: La Galiverna, 1986.

Moore, Patrick. *The Amateur Astronomer's Glossary*. New York: Norton, 1967.

———. *Passion for Astronomy*. New York: Norton, 1991.

Moorman, John. *A History of the Franciscan Order from Its Origins to the*

Year 1517. London: Oxford University Press, 1968; Chicago: Franciscan Herald Press, 1988.

Morgan, Tom. *Saints*. San Francisco: Chronicle, 1994.

Nussdorfer, Laurie. *Civic Politics in the Rome of Urban VIII*. Princeton: Princeton University Press, 1992.

Olson, Roberta J. M. *Fire and Ice: A History of Comets in Art*. New York: Walker, 1985.

Pagano, Sergio M., ed. *I Documenti del Processo di Galileo Galilei*. Vatican City State: Collectanea Archivi Vaticani, 1984.

Paolucci, Antonio, Bruno Pacciani and Rosanna Caterina Proto Pisani. *Il Tesoro di Santa Maria All'Impruneta*. Florence: Becocci, 1987.

Park, Katherine. 'The Criminal and the Saintly Body: Autopsy and Dissection in Renaissance Italy.' *Renaissance Quarterly*, 1994, 1–33.

Pasachoff, Jay M. *Journey Through the Universe*. Orlando: Saunders College Publishing, Harcourt Brace, 1992, 1994.

Pedersen, Olaf. 'Galileo and the Council of Trent: The Galileo Affair Revisited.' In *Essays on the Trial of Galileo*, edited by Richard S. Westfall, 1–43. Vatican Observatory Publications Special Series: *Studi Gallileiani*, 1989.

———. 'Galileo's Religion.' In *The Galileo Affair: A Meeting of Faith and Science*. Edited by G. V. Coyne, M. Heller and J. Zycinski, 75–102. Vatican City State: Specola Vaticana, 1985.

Peterson, Ingrid J. *Clare of Assisi*. Quincy, Ill.: Franciscan Press, 1983.

Pohle, Joseph. *The Sacraments*. Adapted and edited by Arthur Preuss. 4 vols. St Louis: Herder, 1931.

Pulci, Antonia. *Florentine Drama for Convent and Festival*. Annotated and translated by James Wyatt Cook. Chicago: University of Chicago Press, 1996.

Redondi, Pietro. *Galileo Heretic*. Translated by Raymond Rosenthal. Princeton: Princeton University Press, 1987.

Reston, James, Jr. *Galileo, a Life*. New York: HarperCollins, 1994.

Righini Bonelli, Maria Luisa, and Thomas Settle. *The Antique Instruments of the Museum of History of Science in Florence*. Florence: Arnaud, 1973.

Righini Bonelli, Maria Luisa, and William R. Shea. *Galileo's Florentine Residences*. Florence: Istituto di Storia della Scienza, 1979.

Risso, Paolo. *Sulle Orme di Francesco e Chiara.* Torino: Elle Di Ci, 1992.

Rondinelli, Francesco. *Relazione del Contagio stato in Firenze l'anno 1630 e 1633.* Florence: G. B. Landini, 1634, S.A.R. per Jacopo Guiducci, 1714.

Salmon, William. *Salmon's Herbal.* London: I. Dawks, 1710.

Saverio, Francesco, and Maria Rossi. *Galileo Galilei nelle lettere della figlia Suor Maria Celeste.* Lanciano: Rocco Carabba, 1984.

Sella, Domenico. *Italy in the Seventeenth Century.* London: Addison Wesley Longman, 1997.

Sharrat, Michael. *Galileo: Decisive Innovator.* Oxford: Blackwell, 1994.

Shorter Christian Prayer: The Four-Week Psalter of the Liturgy of the Hours. New York: Catholic Book Publishing, 1988.

Viviani Della Robbia, Enrica. *Nei monasteri fiorentini.* Florence: Sansoni, 1946.

Wallace, William A., ed. *Reinterpreting Galileo.* Washington: Catholic University of America Press, 1986.

Ward, J. Neville. *Five for Sorrow, Ten for Joy: A Consideration of the Rosary.* New York: Doubleday, 1973.

Wedgwood, C. V. *The Thirty Years' War.* Garden City: Anchor, 1961.

Westfall, Richard S., ed. *Essays on the Trial of Galileo.* Vatican Observatory Publications Special Series: *Studi Galileiani,* 1989.

Wilkins, Eithne. *The Rose-Garden Game: A Tradition of Beads and Flowers.* New York: Herder and Herder, 1969.

Young, G. F. *The Medici.* 2 vols. New York: Dutton, 1909, 1930.

Ziegler, Philip. *The Black Death.* New York: Harper and Row, 1971.

NOTES

[I] *She who was so precious to you*

p. 6 'I render . . . centuries' is adapted from Albert Van Helden's translation of Galileo's report to the Tuscan court, 30 January 1610 (*Sidereus Nuncius*, pp. 17–18).

p. 8 'I have observed . . . upside down' is Stillman Drake's translation of a letter dated 23 September 1624 (*Galileo at Work*, p. 286).

p. 10 'A woman of exquisite mind . . . to me' comes from Galileo's letter to Elia Diodati, 28 July 1634, translated by Maria Luisa Righini Bonelli and William R. Shea (*Galileo's Florentine Residences*, p. 50).

p. 12 'Whatever the course . . . divine' is taken from Galileo's third letter on sunspots, 1 December 1612, translated by Drake (*Discoveries and Opinions*, p. 128).

[II] *This grand book the universe*

p. 16 'Philosophy . . . labyrinth' is excerpted from Galileo's *The Assayer*, as translated by Drake (*Galileo*, p. 70).

p. 20 'Try, if you can . . . top of the tower' is taken from Drake's translation of the *Dialogue* (p. 223), and 'Imagine them . . . claimed?' is adapted from I. E. Drabkin's translation of 'De motu', as quoted in James MacLachlan, *Galileo Galilei* (p. 24); and 'Aristotle . . . mistake' comes from Drake's translation of *Two New Sciences* (p. 68).

p. 21 The letter beginning 'The present I am going to make Virginia' is translated by Righini Bonelli and Shea (p. 13).

[III] *Bright stars speak of your virtues*

p. 27 'If, Most Serene Prince . . . let alone all' is from Drake's translation of *Operations of the Geometric and Military Compass* (p. 39).

p. 28 'I have waited . . . reflected rays' is from a letter translated by Mario Biagioli in *Galileo, Courtier* (p. 20).

pp. 28–9 'Her Most Serene Highness . . . tomorrow' is translated in Righini Bonelli and Shea (p. 14).

p. 30 'Regarding . . . such a position' is Biagioli's translation (p. 29).

pp. 31–2 All quotes come from *Sidereus Nuncius*. 'And it is like . . . valleys' is Van Helden's translation (p. 40), and 'Planets show . . . a very great deal' is Drake's (*Telescopes, Tides, and Tactics*, p. 49); the sentence fragments on page 33 are from Van Helden (p. 13 and p. 64, respectively).

pp. 33–4 The long passage, 'Your Highness . . . power and authority' is taken from Van Helden (pp. 30–2).

pp. 35–6 Kepler's statement is taken from Van Helden (p. 94).

[IV] *To have the truth seen and recognised*

p. 38 Madonna Giulia's letter translated by Olaf Pedersen in 'Galileo's Religion' (p. 86).

pp. 38–9 Description of Galileo's new house is from Righini Bonelli and Shea (pp. 17–19); as is the letter about his poor health (p. 19).

p. 39 Galileo's description of Saturn from Drake (*Galileo at Work*, p. 163); Kepler's reaction to telescope from I. Bernard Cohen (*Birth of a New Physics*, p. 76). 'In order . . . possible' is from Van Helden (p. 92).

p. 40 Galileo's letter to Salviati, 'I have been . . . gardens, etc.,' translated by Giorgio de Santillana (*Crime of Galileo*, p. 23).

p. 42 Social bulletin translated by Biagioli (p. 253).

pp. 42–3 Lyncean Academy charter is taken from Drake's article in *Science* (p. 1195).

p. 44 Comment on Galileo's debating style is from Biagioli (p. 77), as is Cardinal Barberini's praise (p. 332, n. 89).

pp. 45–6 Letter from Cigoli of 16 December 1611, translated by Drake (*Discoveries and Opinions*, p. 146).

pp. 46–7 Cardinal del Monte's letter from Righini Bonelli and Shea (pp. 20 and 23).

p. 47 Excerpt from *Bodies in Water* translated by Drake (*Cause, Experiment, and Science*, pp. 18–20).

pp. 47–8 Comment on Italian language, 'I wrote . . . them,' from a letter to Paolo Gualdo, translated by Drake (*Discoveries and Opinions*, p. 84).

[V] *In the very face of the Sun*

p. 53 'In that part of the sky . . . brief periods' is from Drake's translation of Galileo's second letter on sunspots (*Discoveries and Opinions*, p. 119).

pp. 53–4 'They wish . . . posterity' is taken from the third letter on sunspots, translated by Drake (*Discoveries and Opinions*, p. 127).

pp. 54–5 Welser's invitation to the discussion is also Drake's translation (*Discoveries and Opinions*, p. 89).

p. 55 Galileo's reference to his indisposition and indecision, 'The difficulty . . . proved it,' is from his first letter on sunspots (*Discoveries and Opinions*, p. 90); and 'With absolute necessity . . . the universe' is also from Drake's translation of the first letter on sunspots (*Discoveries and Opinions*, p. 94).

pp. 55–6 'Sunspots . . . at all' appears further on in Galileo's first letter on sunspots (*Discoveries and Opinions*, p. 98).

p. 56 'I do . . . by us' continues Drake's translation (*Discoveries and Opinions*, p. 100); 'If I may . . . recognise them' is from the first letter on sunspots (*Discoveries and Opinions*, p. 102); 'And forgive me . . . in perfect tune' is from the closing of the first letter on sunspots (*Discoveries and Opinions*, p. 103).

p. 57 Welser's thanks, 'You have . . . lines' and his suggestion, 'It would be . . . however strong,' appears in his second letter to Galileo, translated by Drake (*Discoveries and Opinions*, pp. 104–5).

pp. 57–8, Welser's opening to his second letter, 'I have read . . . Thy sight,' continues Drake's translation (*Discoveries and Opinions*, p. 105).

[VI] *Observant executrix of God's commands*

p. 61 'Thursday morning . . . the telescope' blends two of Drake's translations of Castelli's letter to Galileo (*Galileo at Work*, p. 222, and *Discoveries and Opinions*, p. 151).

pp. 62–3 'After many things . . . that view' and 'Now, getting back . . . never a word' continue Drake's translation of this letter (*Galileo at Work*, pp. 222–3).

p. 63 Galileo's reply to Castelli, 'As to the first . . . of the future,' blends Drake's translation (*Galileo at Work*, pp. 224–5) with others by Olney (quoted in James Brodrick, *Galileo: The Man, His Work, His Misfortunes*, pp. 76–7) and Pedersen (Trent, p. 23).

p. 64 'Holy Scripture . . . God's commands' also combines elements of the translations mentioned immediately above, along with an earlier one by Drake (*Discoveries and Opinions*, p. 182).

p. 65 Continuing Galileo's reply to Castelli, 'I believe . . . so completely,' is again an amalgamation (see Brodrick, pp. 78–9; Drake's *Galileo at Work*, p. 226; and *Discoveries and Opinions*, pp. 183–4).

p. 67 The opening of Galileo's letter to Madama Cristina, 'Some years ago . . . their purposes,' comes from Drake's translation (*Discoveries and Opinions*, p. 175).

pp. 68–70 Continuing the letter, 'Possibly because . . . the Bible,' still from Drake (*Discoveries and Opinions*, p. 177); 'Let us grant . . . his edifices' (*Discoveries and Opinions*, p. 193); and 'To ban Copernicus . . . thousands of years' (*Discoveries and Opinions*, p. 196).

[VII] *The malice of my persecutors*

p. 75 From Galileo's 'Treatise on the Tides,' 'To hold . . . reflections,' translated by Drake, as the passage later appeared in the *Dialogue* (p. 419).

p. 77 Bellarmino's letter combines translations that appear in Jerome J. Langford (*Galileo, Science and the Church*, p. 61), Brodrick (pp. 95–6) and Richard J. Blackwell (*Galileo, Bellarmino, and the Bible*, pp. 265–7).

p. 80 Galileo's letter, 'I told His Holiness . . . on all occasions,' appears in Brodrick (pp. 106–7).

p. 81 Bellarmino's letter supporting Galileo, 'We, Roberto . . . May 1616,' is Sturge's translation quoted in de Santillana's (*Crime of Galileo*, p. 132).

p. 82–3 Galileo's letter to Leopold, 'I send you . . . this chimera,' combines translations by Drake (*Galileo at Work*, p. 262) and de Santillana (p. 151).

[VIII] *Conjecture here among shadows*

p. 86 'As a result . . . by perfect eyes' is from Drake's translation of *The Assayer* (*Controversy*, p. 319; 'I-shall . . . among the graves' is from a letter translated by Righini Bonelli and Shea (p. 19).

p. 89 'During . . . this matter' comes from Drake's translation of *The Assayer* (*Discoveries and Opinions*, p. 236).

p. 92 'Hence . . . imperfect' is taken from Drake's translation of the *Discourse on the Comets* (*Controversy*, p. 57).

p. 93 The gold quip, 'If their . . . my house,' is from Drake's translation of *The Assayer* (in *Discoveries and Opinions*, p. 253, and *Controversy*, p. 229), as is 'That reply . . . duplicity' (*Discoveries and Opinions*, p. 241).

pp. 93–4 'I cannot . . . dray horses' is still from *The Assayer* (*Discoveries and Opinions*, p. 271; *Controversy*, p. 301).

p. 95 From the opening of *The Assayer*, 'I have . . . its intention,' is Drake's (*Discoveries and Opinions*, p. 231).

[IX] *How our father is favoured*

p. 108 Urban's remark about Rome is taken from his bull inaugurating the visitation, quoted in Laurie Nussdorfer (*Civic Politics*, p. 21 and n. 1); the 'ordinary Pope' comment is de Santillana's translation (p. 161).

p. 109 The two short passages from *The Assayer* are both taken from Drake's translation (*Discoveries and Opinions*, p. 239).

pp. 110–11 The parable about the song of the cicada also comes from Drake's translation of *The Assayer* (*Discoveries and Opinions*, pp. 256–8).

[X] *To busy myself in your service*

p. 116 The assessment of Poor Clare life by Maria Domitilla Galluzzi is taken from E. Ann Matter and John Coakley, *Creative Women* (p. 206).

p. 117 Galileo's comment on his 'clown's habit' is recounted by Drake (*Galileo at Work*, p. xiii).

p. 123 Galileo's letter to Castelli is quoted in Righini Bonelli and Shea (p. 32).

[XII] *Because of our zeal*

p. 142 Urban's 'embrace' is quoted in de Santillana (p. 171).

pp. 143–5 The passages from Galileo's 'Reply to Ingoli' are all taken from Maurice A. Finocchiaro (*Galileo Affair*): 'Eight years . . . that time' (p. 154); 'However . . . against my will' (p. 155); 'Note, Signor . . . authority' (p. 155); 'I am thinking . . . faith' (p. 156); 'Thus . . . put together' (p. 156); 'For, Signor . . . the universe' (pp. 156–7); 'If any place . . . therein' (p. 179).

[XIII] *Through my memory of their eloquence*

pp. 147–8 The first quote from the *Dialogue*, 'The constitution . . . works,' is from Drake's translation (pp. 3–4).

p. 149 The excerpt from the *Dialogue*'s preface, 'Many years . . . specu-lations,' is Finocchiaro's translation (*World Systems*, p. 81).

pp. 149–50 'Now, since . . . reflections' is a mix of Drake and Finocchiaro (*Dialogue*, p. 7, and *World Systems*, p. 82).

p. 150 The comment on the diagrams is from Drake (*Dialogue*, p. 80).

pp. 150–1 'Some years . . . prohibitions' is Drake's translation (*Dialogue*, p. 5).

p. 151 'Upon hearing . . . mind as well' continues Drake's translation of the *Dialogue*'s preface (p. 5).

p. 152 'For my part . . . nonexistent' is from Drake (*Dialogue*, pp. 58–9).

pp. 152–3 'The deeper . . . they are' (Drake's *Dialogue*, p. 59).

p. 154 Galileo's critique of the invention is Drake's translation (*Galileo at Work*, p. 297).

[XIV] *A small and trifling body*

All quotations from the *Dialogue* in this chapter come from Drake's translation.

p. 160 'I act . . . the stage' (p. 131).

p. 161 'The air . . . forever asleep' (p. 183); and 'Shut . . . standing still' (pp. 186–7).

p. 162 'We encounter . . . violence' (p. 120).

p. 163 Michelangelo's letter is from Righini Bonelli and Shea (p. 32).

[XV] *On the right path, by the grace of God*

pp. 175–6 Galileo's letter to Elia Diodati is Drake's translation (*Galileo at Work*, p. 310).

All the excerpts from the *Dialogue* in this chapter are Drake's translations:

p. 177 'But another effect . . . mighty marvel' (p. 345).

p. 178 'I might . . . unbounded' (p. 319).

p. 178–9 'It seems to . . . serve us' (p. 368).

p. 179–80 'I believe . . . corpses' (p. 368).

p. 180–1 'Besides . . . mankind' (pp. 368–9).

[XVI] *The tempest of our many torments*

Again drawing on Drake's translation of the *Dialogue* for the quotations in this chapter:

p. 184 The anecdote about Aristotle is from Day Four (p. 433).

p. 185 'In the . . . tides' (p. 462).

p. 185–6 'As to . . . his own' (p. 464).

[XVII] *While seeking to immortalise your fame*

p. 198 Castelli's letter is quoted in Pedersen ('Religion', p. 94).

p. 199 Father Riccardi's imprimatur for *The Assayer* is Drake's translation (*Controversy*, p. 152).

pp. 203–204 Galileo's letter (to Baliani, 6 August 1630) is Drake's translation (*Galileo at Work*, p. 313).

[XVIII] *Since the Lord chastises us with these whips*

p. 209 Petrarca's comment on 'happy posterity' is quoted from *Epistolae Familiaris* in Philip Ziegler, *The Black Death* (p. 45).

[XX] *That I should be begged to publish such a work*

All the official correspondence in this chapter is translated by Finocchiaro in *The Galileo Affair*.

p. 227 The sentence fragment describing the ideas in the *Dialogue* as 'chimeras, dreams,' is drawn from Galileo's letter to the Tuscan secretary of state, 7 March 1631 (*Galileo Affair*, p. 207); two more excerpts from this letter: 'Indeed . . . such a work' (*Galileo Affair*, p. 207–8) and 'In the meantime . . . ill health' (*Galileo Affair*, p. 208).

pp. 227–8 Father Riccardi's letter of 24 May, 'I want . . . revised' (*Galileo Affair*, p. 212).

p. 228 Father Riccardi's 19 July letter, 'In . . . content' (*Galileo Affair*, p. 213); his instructions for the ending (*Galileo Affair*, p. 354, n. 57).

p. 229 Ambassador Niccolini's letter, 'After . . . Most Serene House' (*Galileo Affair*, p. 214).

p. 229 Galileo's dedication to Ferdinando II, 'These dialogues . . . publication', is Drake's translation (*Dialogue*, pp. 3–4).

p. 232 Castelli's letter (29 May 1632), 'I still . . . to myself' is translated by Drake (*Galileo at Work*, pp. 336–37).

p. 234 The commissioners' report to the pope, 'We think . . . printed book,' is Finocchiaro's translation (*Galileo Affair*, p. 219); Ambassador Niccolini's views of the pope: 'I feel . . . rage' (*Galileo Affair*, p. 229), and 'When . . . a troublesome affair' (*Galileo Affair*, pp. 231–2).

pp. 235–6 Galileo's letter to Diodati of 15 January 1633 is de Santillana's translation (pp. 215–16 n. 18).

[XXI] *How anxiously I live, awaiting word from you*

All of the diplomatic correspondence in this chapter is translated by Finocchiaro, and all page numbers in parentheses refer to *The Galileo Affair*.

p. 243 Niccolini's reference to the secrecy of the Holy Office (p. 240).

p. 244 Niccolini's report on the first week, 'The latter . . . to him' (p. 244).

pp. 249–50 Niccolini's dispatch of 6 March, 'About . . . the matter' (p. 246).

p. 251 Niccolini's letter, 'I reiterated . . . these subjects' (p. 247).

[XXII] *In the chambers of the Holy Office of the Inquisition*

The trial transcript, which appears in this chapter in its entirety, is drawn mostly from Finocchiaro's translation published in *The Galileo Affair*. Several passages, however, are blended with the partial translation by Drake in *Galileo at Work*, and the whole is informed by de Santillana's treatment of the transcript in *The Crime of Galileo*.

pp. 255–60 The first deposition combines Drake (pp. 344–7), Finocchiaro (pp. 256–62), and de Santillana (pp. 237–40). The first excerpt from the Inquisition's dossier, 'His Holiness . . . imprisoned' (Finocchiaro, p. 247, and de Santillana, pp. 125–56) The 26 February entry, 'In the Palace . . . against him' (Drake, p. 348).

pp. 261–4 The continuation of the first deposition is a mix of Drake (p. 347), as well as of another partial translation in Langford (p. 139), and Finocchiaro (pp. 260–2).

p. 265 The close of the first (12 April 1633) deposition (Finocchiaro, p. 262).

[XXIII] *Vainglorious ambition, pure ignorance and inadvertence*

p. 268 Inchofer's statements on the *Dialogue* are Finocchiaro's translations (*Galileo Affair*, p. 264 and p. 266); and the continuation of the statements, 'If Galileo . . . in mind,' is a mix of Finocchiaro (*Galileo Affair*, p. 268) and de Santillana (pp. 246–7).

pp. 268–9 The excerpts from the Father Commissary's letter are taken from Drake's translation (*Galileo at Work*, pp. 349–50).

pp. 269–70 Galileo's second deposition (30 April) remarks are a blend of de Santillana (pp. 255–6), Langford (pp. 144–5), and Finocchiaro (*Galileo Affair*, p. 278).

p. 270 Niccolini's observation, 'It is . . . alive,' is de Santillana's translation (p. 258).

pp. 273–4 The excerpts from Galileo's written defence statement are drawn from Finocchiaro's translation (*Galileo Affair*, pp. 279–81) with a few minor editorial changes.

p. 274 The quote 'Lastly . . . prospect' contains some part of F. S. Taylor's translation as cited in Langford (p. 147).

p. 275 Niccolini's letter, 'In regard . . . myself,' is Finocchiaro's translation (*Galileo Affair*, p. 253).

[XXIV] *Faith vested in the miraculous Madonna of Impruneta*

pp. 283–5 Galileo's deposition of 21 June 1633, is a mix of Finocchiaro (*Galileo Affair*, pp. 286–7) and de Santillana (pp. 302–3), with two minor editorial adjustments.

[XXV] *Judgment passed on your book and your person*

p. 288 Galileo's sentence is a blend of Langford (pp. 152–3) and the text posted on the Website of the Istituto e Museo della Storia di Scienza (galileo.imss.firenze.it).

pp. 289–91 The text of Galileo's abjuration is a combination of de Santillana (pp. 312–13) and Righini Bonelli and Shea (pp. 48–9).

[XXVI] *Not knowing how to refuse him the keys*

p. 300 Piccolomini's letter to Galileo is from de Santillana's translation (p. 200).

p. 305 Galileo's letter (to Nicolas-Claude Fabri de Peiresc) is de Santillana's translation (p. 324).

[XXVII] *Terrible destruction on the feast of San Lorenzo*

p. 310 The archbishop's assessment of the bell casting is from Drake's translation (*Galileo at Work*, p. 355).

p. 315 The quote about 'MOTION' (p. 147) – and all other excerpts from *Two New Sciences* in this chapter – are drawn from Drake's translation.

pp. 315–16, 'Just as . . . his books' (p. xiii); 'There will . . . still deeper' (p. 147).

[XXVIII] *Recitation of the penitential psalms*

p. 322 'The constant . . . mechanics' is from Henry Crew and Alfonso de Salvio's translation of *Dialogues Concerning Two New Sciences* (p. 1).

p. 323 Sagredo's comments on large structures are a mix of Crew and de Salvio (p. 1) and Drake (*Two New Sciences*, p. 11).

p. 323 Salviati's response, 'Please observe . . . manifest error', is mostly from Crew and de Salvio (pp. 4–5), with measurement figures from Drake.

p. 324 Simplicio's satisfaction is quoted from Drake (*Two New Sciences*, p. 93). 'To illustrate . . . own size' is from Crew and de Salvio (p. 131).

p. 329 Excerpt from Galileo's letter to Peiresc is translated by Mary Allan-Olney (*Private Life of Galileo*, pp. 278–9) and cited in Drake's *Galileo* (pp. 92–3).

[XXIX] *The book of life, or, A prophet accepted in his own land*

pp. 339–40 Galileo's defence of Girolamo Fabrici of Acquapendente is from Drake's translation (*Galileo at Work*, pp. 172–3), with minor changes.

p. 341 The anonymous denunciation of the archbishop is de Santillana's translation (p. 325, n.4), with minor changes based on my reading of the original in Pagano.

[XXX] *My soul and its longing*

Most of the quotes from *Two New Sciences* in this chapter come from Crew and de Salvio:

p. 348 Salviati's description of the ball-rolling trials (pp. 178–9).

pp. 348–9 'For the . . . results' (p. 179).

p. 350 'One cannot . . . curve' (p. 250); and 'Your demonstration . . . from true' (p. 248).

p. 351 'The force . . . repeated experiment' (p. 276); 'The cause . . . investigation' actually comes from Kline, *Mathematical Thought* (p. 333), but resembles Crew and de Salvio (p. 166).

[XXXI] *Until I have this from your lips*

pp. 358–9 Galileo's letter to Cardinal Barberini of 17 December 1633, is my own translation.

p. 359 Aggiunti's letter is translated by Pedersen ('Religion', p. 88), with slight modifications.

pp. 360–1 The condolence letters from the ambassadress, the archbishop and Signor Geri are my own translations.

p. 361 'I feel . . . to me' is Drake's translation (*Galileo at Work*, p. 360); while 'I do . . . afraid' is Pedersen's ('Religion', p. 88).

pp. 361–2 Galileo's letter to Diodati is a blend of Righini Bonelli and Shea (p. 50) and de Santillana (p. 223).

[XXXII] *As I struggle to understand*

p. 363 'The treatise . . . there' is from Drake's translation (*Galileo at Work*, p. 362).

p. 364 'I find . . . younger' is a mix of Drake (*Galileo at Work*, p. 375) and Sharratt (p. 185); 'You have . . . all sides' is de Santillana's translation (p. 324) of Galileo's letter to Peiresc, 16 March 1635.

p. 366 Fra Micanzio's comment on transcription is from Drake (*Galileo at Work*, p. 382); and 'I shall . . . not be lost' is also from Drake (*Galileo at Work*, p. 375).

pp. 367–8 Galileo's preface to *Two New Sciences* is taken from Drake's translation (pp. 5–6).

pp. 368–9 The excerpts from letters of the pope's brother to the Florentine inquisitor are Pedersen's translations ('Religion', p. 100).

p. 370 Galileo's comments on the Moon's libration are translated by Drake (*Galileo at Work*, p. 385), with a few word substitutions.

p. 371 'This universe . . . my body' is my translation.

[XXXIII] *The memory of the sweetnesses*

p. 375 Vincenzio's recollections of his father are translated in Francesco Bertola (*Da Galileo alle Stelle*, p. 101).

pp. 375–6 'The falsity . . . easily discovered' is from Drake's translation (*Galileo at Work*, p. 417).

pp. 376–7 'Bereft . . . with you' is Pedersen's translation ('Religion', p. 83).

p. 377 'I have . . . pilgrim minds' is from Drake's translation (*Galileo at Work*, p. 397), as is 'I hope . . . of mine' (*Galileo at Work*, p. 421).

p. 378 Lucas Holste's eulogy is Drake's translation (*Galileo at Work*, p. 436, and *Galileo*, p. 93).

APPRECIATION

I most sincerely thank Silvio Bedini for bringing Suor Maria Celeste into my life; Albert Van Helden for encouraging me to tell her story; George Gibson for wanting to hear it; Michael Carlisle for retrieving a treasure from Venice; Kristine Puopolo for her curiosity; John Casey for his clues; Father Ernan McMullin for his insights; Mariarosa Gamba Frybergh and Alfonso Triggiani for the Italian lessons; I. Bernard Cohen for his blessing and dazzling; Doron Weber and the Alfred P. Sloan Foundation for the officers' grant; William J. H. Andrewes for his support; Betty Sobel for her research assistance; Owen Gingerich for his challenge and the view from the Geniculum; Stephen Sobel for the lute music and calendrics; Robert Pirie and the American Academy in Rome for the night of the Lynx; Ken Soden and Frank Randazzo for the itineraries; Irene Tully for the poem; Drs Michael and Stephen Sobel, Peter Michalos, Barry Gruber, Alan Katz and Harry Fritts for their diagnoses of diseases past; Flanzy Chodkowski for the textbooks, hagiographies and rosaries; Diane Ackerman and Lois Morris for the notebooks; Antonia Ida Fontana and the National Central Library of Florence for permission to view Suor Maria Celeste's letters; Franco Pacini for the keys to Galileo's house; Paolo Zaninoni for source materials from Italy; Mara Miniati for *carte blanche* at the Museum of the History of Science in Florence; Paolo Galluzzi for the secret of Galileo's tomb; Francesco Bertola for arriving at Padua *deus ex machina*; Frank Drake for his celestial mechanics; Chiara Peacock and Barbara Lynn-Davis for the Tuscan gardens; Antonio Di Nunzio for entry into the Clarisse convents of Torino; Amanda Sobel for interstate library loans; James MacLachlan for his work in progress and Mersenne tradition of generosity; K. C. Cole for her wisdom; Kate Epstein for her Latin erudition; Mother Mary Francis and her sisters at the Poor Clare Monastery of Our Lady of Guadalupe for their prayers and the answers to my questions; Thomas Settle for conducting experiments in the history of science; the staff members of the book departments at Christie's and Sotheby's New York auction houses and Betsy Walsh at the Folger Shakespeare Library in Washington for access to first editions of Galileo's books; Marcy Posner and Tracy Fisher for representation in foreign markets; Rita and Gary Reiswig for the festivities; and Zoë and Isaac Klein for their love, support, finger puppets and inspirational icons.

Additional thanks to Bernard Cohen, Frank Drake, Mariarosa Frybergh, Owen Gingerich, James MacLachlan, Mother Mary Francis, Christopher Potter, Dick Teresi, Alfonso Triggiani and Albert van Helden for reviewing and commenting on the almost-final manuscript.

PICTURE CREDITS

Page 187 ISTITUTO E MUSEO DI STORIA DELLA SCIENZA, Florence
Page 192 The Senators of Florence, by Justus Sustermans. ASHMOLEAN
          MUSEUM, Oxford
Page 196 THE TIME MUSEUM, Rockford, Illinois
Page 203 BIBLIOTECA NAZIONALE, Florence
Page 211 NIMATALLAH / ART RESOURCE, NY
Page 212 THE HOUGHTON LIBRARY, Harvard University, Cambridge,
          Massachusetts
Page 223 SCALA / ART RESOURCE, NY
Page 224 ISTITUTO E MUSEO DI STORIA DELLA SCIENZA, Florence
Page 226 BIBLIOTECA APOSTOLICIA VATICANA, Rome
Page 231 ISTITUTO E MUSEO DI STORIA DELLA SCIENZA, Florence
Page 238 Trial of Galileo. Anonymous. THE BRIDGEMAN ART LIBRARY
Page 243 THE FOLGER SHAKESPEARE LIBRARY, Washington, D.C.
Page 244 BIBLIOTECA NAZIONALE, Florence
Page 263 MUSEE DU LOUVRE, Paris
Page 269 THE LIBRARY OF CONGRESS
Page 280 IL MUSEO DEL TESORO DI SANTA MARIA ALL-IMPRUNETA
Page 291 ARCHIVIO SEGRETO VATICANO
Page 296 Galileo and his incline plane. ARCHIVI ALINARI / ART RESOURCE, NY
Page 301 BIBLIOTHEQUE NATIONALE, Paris
Page 316 INSTITUTO E MUSEO E MUSEO DI STORIA DELLA SCIENZA
Page 322 Christie's Images, London THE BRIDGEMAN ART LIBRARY
Page 324 From *Two New Sciences*, by Galileo Galilei, translated by Henry Crew
          & Alphonso de Salvio, Dover Publications, New York
Page 339 IMAGE SELECT / ART RESOURCE, NY
Page 342 Galileo and Vincenzio Viviani. INSTITUTO E MUSEO DI STORIA DELLA
          SCIENZA, Florence
Page 346 ART RESOURCE, NY
Page 349 INSTITUTO E MUSEO DI STORIA DELLA SCIENZA, Florence
Page 359 FRATELLI ALINARI / ART RESOURCE, NY
Page 365 INSTITUTO E MUSEO DI STORIA DELLA SCIENZA, Florence
Page 367 INSTITUTO E MUSEO DI STORIA DELLA SCIENZA, Florence
Page 374 INSTITUTO E MUSEO DI STORIA DELLA SCIENZA, Florence
Page 376 SCALA / ART RESOURCE, NY
Page 381 ART RESOURCE, NY
Page 383 INSTITUTO E MUSEO DI STORIA DELLA SCIENZA, Florence

INDEX

LONGITUDE
Dava Sobel

This phenomenal number one bestseller is the dramatic, true story of John Harrison's epic quest to solve the greatest scientific dilemma of his time: the longitude problem.

'Rarely have I enjoyed a book as much as Dava Sobel's *Longitude*. She has an extraordinary gift of making difficult ideas clear.' *Daily Telegraph*

1 85702 571 7   £5.99

THE ILLUSTRATED LONGITUDE
Dava Sobel

With new material from Dava Sobel and William Andrewes, and illustrated with over 200 integrated photographs, *The Illustrated Longitude* is the essential book for everyone who fell in love with John Harrison's story and wants to know more.

1 84115 233 1   £14.99

FERMAT'S LAST THEOREM
Simon Singh

The extraordinary story of a riddle that confounded the world's greatest minds for 358 years, and how an Englishman, after years of secret toil, finally solved mathematics' most challenging problem.

'A magnificent story, if you enjoyed *Longitude* you will enjoy this.' *Evening Standard*

1 85702 669 1   £6.99

THE CODE BOOK

The Secret History of Codes and Code-breaking

Simon Singh

'You couldn't wish for a better guide than Singh. As he demonstrated in his last book, *Fermat's Last Theorem*, he has the priceless knack of being able to strip away jargon and describe mathematical processes in something so akin to clear English that even your average non-scientific duffer is able to convince himself that he understands it.' *Mail on Sunday*

1 85702 889 9   £7.99

ISAAC NEWTON: The Last Sorcerer

Michael White

A new biography of Isaac Newton that reveals the extraordinary influence that the study of alchemy had on the greatest Early Modern scientific discoveries. In this 'ground-breaking biography' Michael White destroys the myths of the life of Isaac Newton and reveals a portrait of the scientist as the last sorcerer.

'A ground-breaking biography.' *Sunday Times*

1 85702 706 X   £8.99

THE MAN WHO LOVED ONLY NUMBERS

The Story of Paul Erdos and the Search for Mathematical Truth

Paul Hoffman

The biography of a mathematical genius. Paul Erdos was the most prolific pure mathematician in history and, arguably, the strangest too.

'Hoffman's playful, plainspoken and often hilarious biography of a monkish, impish, generous genius is purest pleasure.' *Mail on Sunday*

1 85702 829 5   £7.99